BASIC INTEGRATED CIRCUIT THEORY & PROJECTS

No. 1699
$19.95

BASIC INTEGRATED CIRCUIT THEORY & PROJECTS

BY CHARLES K. ADAMS

TAB BOOKS Inc.
BLUE RIDGE SUMMIT, PA. 17214

FIRST EDITION

FIRST PRINTING

Printed in the United States of America

Library of Congress Cataloging in Publication Data

Adams, Charles K.
Basic integrated circuit theory and projects.

Includes index.
1. Integrated circuits. I. Title.
TK7874.A33 1984 621.381'73 84-8730
ISBN 0-8306-0699-8
ISBN 0-8306-1699-3 (pbk.)

Cover photograph courtesy of Motorola Semiconductor Products Inc.

Contents

Introduction

The purpose of this book is to help you understand how an integrated circuit works and how they are used. Both analog and digital integrated circuits are discussed. First, semiconductor theory is covered in simplified terms to give you an understanding of how semiconducotrs work. Then the fabrication of transistors and integrated circuits is discussed to give you an insight into the manufacturing methods required.

Next, the functions of some of the ICs are discussed in detail. Then the specifications are covered for the representative ICs. The typical applications for both the analog and digital ICs are discussed next. The final chapter covers some projects using the integrated circuits that have been studied. These projects are for you to assemble and use. They will familiarize you with ICs.

This book will give you working knowledge of integrated circuits and their use. It is not a rigorous treatment of integrated circuits, but it will give you the basics so that integrated circuits are not a complete mystery.

Read it one chapter at a time, and then study the chapter until the information presented is clearly understood. It will be hard to understand everything about all of the devices the first time through. In the chapters covering the specific devices, study a few of the devices at a time. Understand their operation and function before moving on. Take your time while building the projects. If the circuit is not fully understood, review the information given on the device.

I assume that you have some basic knowledge of electronics and that you know the basic circuit elements, such as resistors and capacitors. It is also assumed that you have an interest in electronics and enjoy building circuits that work. After assembling the projects, you will be able to use ICs in other circuits and even design your own simple IC circuits. The electronics field is varied and complex. This exposure to a small portion of it does not make you an expert, but with this book as a starting place you can gain the background you need to move on to more advanced books.

Acknowledgments

I would like to express sincere thanks to the following companies for allowing the use of their information in this book.

Fairchild Camera and Instrument Corporation, Linear Division
National Semiconductor Corp.
Motorola Semiconductor Products Inc.

1

Miniaturization

The electronics revolution is here and in full swing. The new electronic devices have, in some manner, touched all of us—whether it be as electronic games, digital watches, electronic banking, satellite communications, computers, or any of the other fantastic devices made possible by the advances in the electronic field. It was in the early 1950s that the forerunner for this revolution was born. The transistor was the first widely used semiconductor device. Since then, new developments have appeared which could not even be conceived at that time. These developments have touched on all fields in fantastic ways. Several jobs have been made easier, such as chemical analysis and banking. Several others have been made possible by the new electronic devices, such as satellites, spacecraft control, and computers.

With the advent of the transistor, it is no longer necessary to boil off electrons from a hot filament to produce an electron flow that can be controlled. The transistor made it possible to control the electron flow in a solid piece of material, the semiconductor. The vacuum tube generated heat, required different power supplies for the filament and plate voltages, and was relatively large. The transistor was small, used only one low voltage, and required less power. Consequently, less heat was generated. Figure 1-1 shows the size comparison between tubes and transistors.

The early transistors were simple devices, with characteristics that are outdated by today's transistors. Table 1-1 compares

Table 1-1. Comparison of Transistors

Number	Type	Vc (volts)	Ic (mA)	Pd (mW)	Gain	Freq (MHz)
CK722	pnp	22	10	180	45	0.8
2N34	pnp	25	8	50	40	0.6
2N35	npn	25	8	50	40	0.8
2N108	pnp	20	15	58		
2N180	pnp	30	25	150	60	0.7
2N708	npn	40		360	50	500
2N3568	npn	80	100	125	80	60
2N3704	npn	50	80	360	150	100
2N3906	pnp	40		300	200	250

some of the early transistor specifications with those of some current transistors. The early devices were used by both the hobbyist and the designer. They paved the way for the developments which led to modern transistors. These early transistors replaced tubes in many applications with the advantages of requiring only one lower voltage and smaller size. In many instances, several transistor circuits could be placed in the same space that a single tube circuit occupied.

In many cases, the early transistor circuits were simply tube circuits adapted for transistors. Before long transistor circuits were being designed from scratch. This resulted in the optimization of

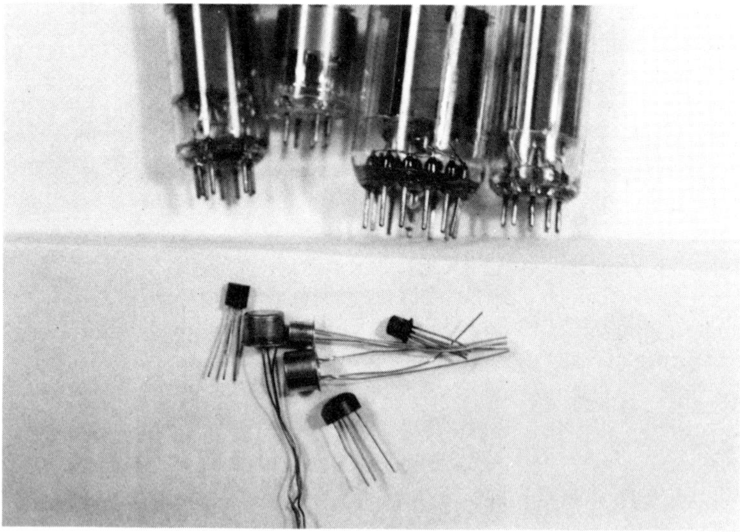

Fig. 1-1. Photograph of some tubes and transistors to illustrate the size difference.

the transistor parameters, which led to better and more efficient transistors.

Semiconductor theory, which explains how a solid semiconductor material can conduct and control current flow, required a new way of thinking. This theory delves into the atomic structure of crystalline material and the behavior of individual atoms in a crystal lattice.

The advent of the space program financed some of the major advances in electronics. The space program required more efficient and more reliable devices. The study of these factors led to better manufacturing methods and improved technology, and the need for more capability in less space brought about the revolution in miniaturization. As the complexity of the spacecraft increased, the technology kept pace, giving better and better semiconductor devices.

The transistor is one active element. It was logical that the transistor would be combined with some passive elements, such as resistors, capacitors, and diodes, to achieve a functional element. Thus, the first integrated circuit was born. To make the resistors on the same semiconductor substrate as the transistor required advancing the technology to control the resistance. These early integrated circuits were simple equivalents of transistor circuits. As

Fig. 1-2. Photograph of some 16-pin and 40-pin integrated circuits to illustrate the size difference.

the knowledge of how to make integrated circuits increased and the equipment became available to fabricate more complex devices, integrated circuits were soon made with several transistors and other components. Presently, there are integrated circuits, or chips, with thousands of transistors. Complex integrated circuits can now replace many less complex integrated circuits.

□ Clock chips allow simple, inexpensive digital clocks and watches.
□ Calculator chips make the inexpensive calculator available in many forms.
□ Video game chips give rise to the popularity of video games.
□ Microcomputer chips lead the way to the popularity of the microcomputer.

Figure 1-2 shows some of the larger and more complex integrated circuits, along with some of the smaller and simpler ICs.

During this same time, analog integrated circuits began to appear. They were slower in developing than the digital integrated circuits because they required more precise components for proper control of the analog levels. The problems were overcome and they soon became commonplace. Presently, several IC manufacturers are advancing the state of the art for analog integrated circuits. In a few years, the complexity and versatility of analog ICs will match those for digital circuits.

2

Semiconductor Theory and Manufacturing

This chapter covers the semiconductor theory in a very simplified form, and the manufacturing methods for semiconductor devices and transistors. It takes this process up to the manufacturing of the integrated circuit, which is covered in the next chapter.

Many single semiconductor devices are developed from the common starting point, the pn junction. These include single junction devices such as the diode and varistor, and devices with two or more junctions, such as the transistor, field effect transistor, silicon controlled rectifier, and unijunction transistor. Integrated circuits fall in the category of multicomponent devices.

Single junction devices are usually passive devices, with the exception of the tunnel diode. They provide no signal gain. Devices with two or more junctions are active devices, providing current and/or voltage gain.

ATOMIC STRUCTURE

The *atom* is a basic building block of nature. All things are made up of them. It is the atoms of the material and how they are combined with other atoms that defines what the material is and the properties it has. It can be air, rock, food, water, or whatever. A cursory look at the atom is required as a starting point for semiconductor theory.

An atom consists of a *nucleus* with *electrons* in orbit around it.

Table 2-1. A Portion of the Periodic Table

Period	Groups					Electrons To Fill Orbit
	Ib	IIb	IIIb	IVb	Vb	
3			Al 13	Si 14	P 15	8
4	Cu 29	Zn 30	Ga 31	Ge 32	As 33	8
5	Ag 47	Cd 48	In 49	Sn 50	Sb 51	8
6	Au 79	Hg 80	Tl 81	Pb 82	Bi 83	8

The nucleus contains *protons* and *neutrons.* The protons have a positive charge and are equal in number to the electrons in orbit. The electrons occupy orbits around the nucleus. The orbits are fixed and only a certain number of electrons can occupy each orbit.

Table 2-1 shows a part of the *periodic table of elements* that contains the semiconductors. The periodic table of elements defines the atomic weight, atomic number, number of orbits for electrons, and the number of electrons for all known elements. Table 2-1 shows groups 1b through 5b and periods 3 through 6 and gives the chemical symbol and the atomic number for the element. Table 2-2 gives the chemical name for the symbols shown in Table 2-1.

The *group number* refers to the number of electrons in the outer orbit of the atom. The *atomic number* refers to the total number of electrons (or protons). A standard representation of an atom of aluminum is shown in Fig. 2-1. From the periodic table, it is seen that aluminum (Al) has an atomic number of 13 and is in group

Table 2-2. The Chemical Names for the Element Symbols

Symbol	Element	Symbol	Element
Al	Aluminum	Hg	Mercury
Ag	Silver	In	Indium
Au	Gold	P	Phosphorus
As	Arsenic	Pb	Lead
Bi	Bismuth	Sb	Antimony
Cd	Cadmium	Si	Silicon
Cu	Copper	Sn	Tin
Ga	Gallium	Tl	Thallium
Ge	Germanium	Zn	Zinc

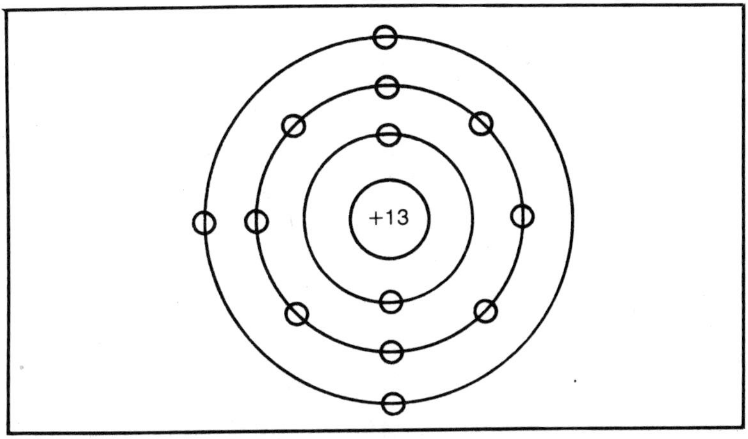

Fig. 2-1. Atomic diagram for aluminum showing the nucleus and the electrons in their orbits.

3b. These numbers tell you that there are 13 electrons in orbit with three in the outer orbit. The nucleus has 13 protons with the nucleus charge shown in the center circle. The number of electron orbits is given by the *period* of the element. For aluminum, the period is three, and so there are three different orbits of electrons. Each inner orbit must be filled before another orbit can be called, and each orbit contains a definite maximum number of electrons. Table 2-3 gives the number of electrons that the various orbits can hold for periods 1 through 6. Note that an outer orbit can never hold more than eight electrons and that the inner orbit can hold up to 2 electrons.

The electrons in the outer orbit are called *valence electrons. Valence* is defined as the number of hydrogen atoms which can combine with one atom of the element. Hydrogen has one electron and an atomic number of 1. Since the maximum number of electrons for the outer orbits is eight, the valence is the number of electrons required to fill this orbit to eight. The aluminum atom has three electrons in its outer orbit. Five more electrons are needed to reach

Period	Electrons in Orbits
1	2
2	2_8
3	2_8_8
4	2_8_18_8
5	2_8_18_18_8
6	2_8_18_32_18_8

Table 2-3. The Electron Capacity of the Orbits

eight. Therefore, it has a valence of 5, and it will combine with five hydrogen atoms. This is illustrated in Fig. 2-2. The electrons which occupy places in the orbits of both the hydrogen and aluminum atoms are called *shared electrons*. That is, they are shared between two atoms. In Fig. 2-2, the atoms with a +1 in the nucleus and one electron in orbit are the hydrogen atoms. The other atom with the +13 in the nucleus is the aluminum atom.

It is the electrons in the outer orbit which combine with the electrons in the outer orbits of other atoms to form compounds. How they combine, and with what elements is very complex and will not be covered here. But it is these electrons which are responsible for most of the characteristics of an element.

SEMICONDUCTOR MATERIAL

The resistance an element shows to electron (or current) flow is the resistivity. For current to flow, electrons must be able to be replaced to form free electron flow through the material. Figure 2-3 shows a resistivity scale for some of the common materials. The

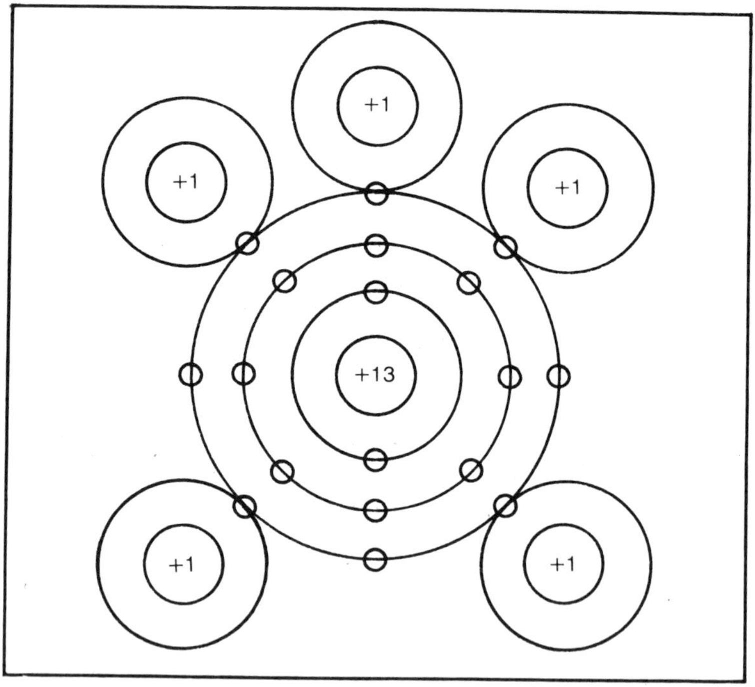

Fig. 2-2. One aluminum atom combined with five hydrogen atoms showing the shared electrons.

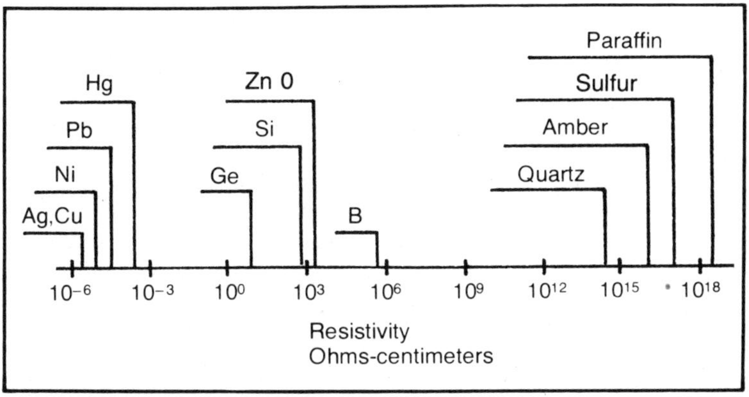

Fig. 2-3. Resistivity scale.

lower resistivity represents conductors, or those able to conduct electricity, or able to sustain current flow, with little loss. Those with the high resistivity, on the right side of the figure, are insulators. They offer a high resistance to electron flow. Those in the center are semiconductors. The resistivity is too high to be considered conductors and too low to be considered insulators. It is these elements, mainly germanium and silicon, which are used for semiconductor devices. They have a valence of 4.

Germanium, of Ge, has an atomic number of 32, and a valence of 4. It is in group 4b of the periodic table with a period of 4.

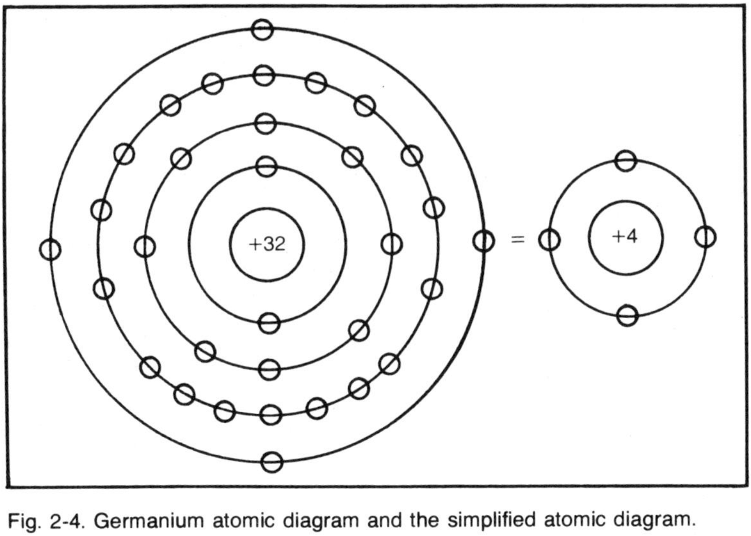

Fig. 2-4. Germanium atomic diagram and the simplified atomic diagram.

Therefore, you know that there are four orbits of electrons with 2, 8, 18, and 4 electrons in the orbits progressing out from the nucleus (refer to Table 2-3). The atomic diagram is given in Fig. 2-4 along with the simplified diagram.

The full atomic diagram is complex, so it is usually represented by the simplified diagram. The inner orbits of the atom are not important in the understanding of how current flow occurs so they are usually left out. In the simplified diagram, the atom is represented by an inner circle and the outer electron orbit. The inner circle represents the nucleus and all the inner orbits. Therefore, it will have a charge equal to the protons which are positive and the electrons represented, which are negative ($+32 - 28 = 4$). The +4 represents the number of electrons in the outer orbit. Realize that the simplified diagram is just that and does not contain the atomic number.

The other popular semiconductor material is silicon, or Si. Figure 2-5 shows the atomic diagram for silicon along with the simplified diagram. It has an atomic number of 14 and a valence of 4. There are two electrons in the inner orbit, eight electrons in the next orbit, and four electrons in the outer orbit. This corresponds to the atomic number of 14. Notice that the simplified diagram for silicon is the same as that for germanium.

Crystals of these materials are formed when several atoms share common electrons to fill the outer orbits. Since this combination results in all the outer orbits being full, there are no outer

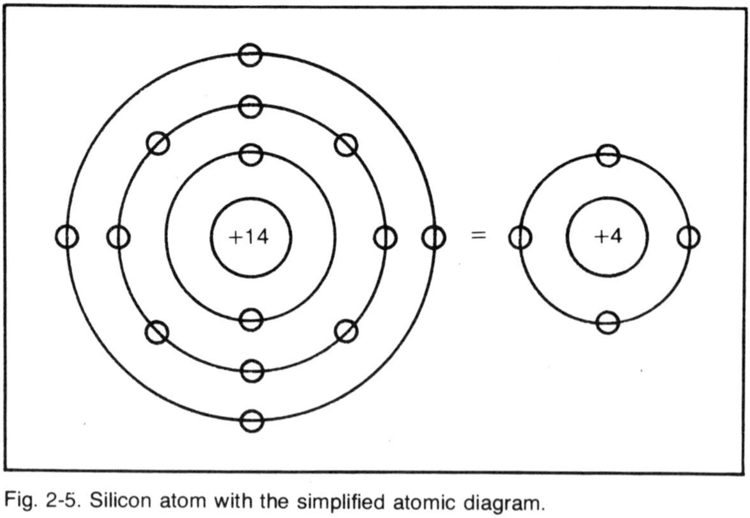

Fig. 2-5. Silicon atom with the simplified atomic diagram.

electrons that are not shared by two atoms. Therefore to sustain current flow considerable energy is required to release the required electrons from the atomic bonds. Figure 2-6 shows a section of a semiconductor crystal. All of the electrons are shared between two atoms.

An atom with a valence of 3 has five electrons in the outer orbit. The valence of 3 means that it takes three more electrons to fill the outer orbit. An atom with the valence of 5 has three electrons in the outer orbit. Therefore it takes five electrons to fill this orbit.

If one atom with a valence of 3 is added to a semiconductor crystal, as shown in Fig. 2-7, there is one electron which is not shared by two atoms. Therefore it takes less energy to free this electron from the parent atom. When this electron is freed, the atom takes a positive charge. This is because there are five protons in the nucleus, and only four electrons in the outer orbit. If an electric field is applied across the crystal, the extra electron will be drawn toward the positive side of the field. This atom is able to donate an electron to help sustain current flow; therefore, it is called a *donor.*

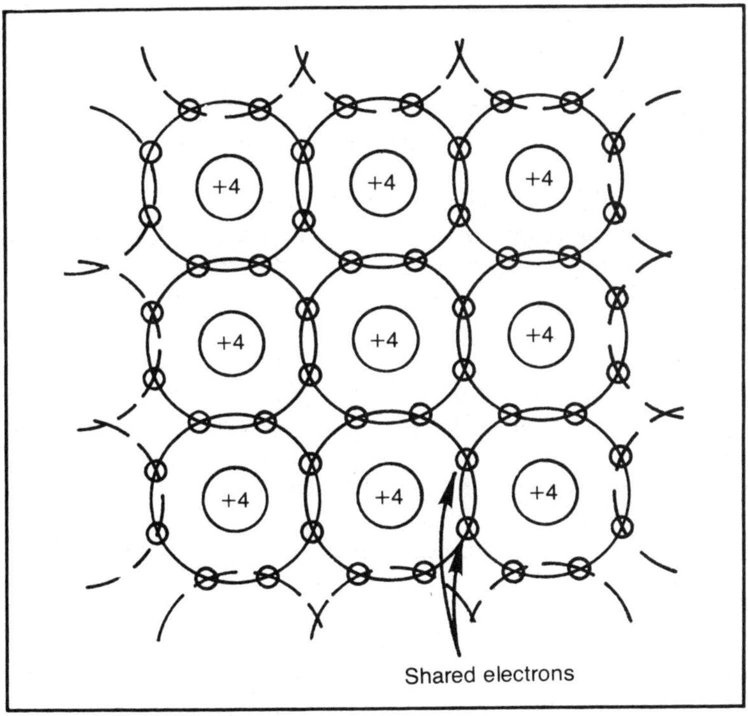

Fig. 2-6. Semiconductor crystalline structure atomic diagram.

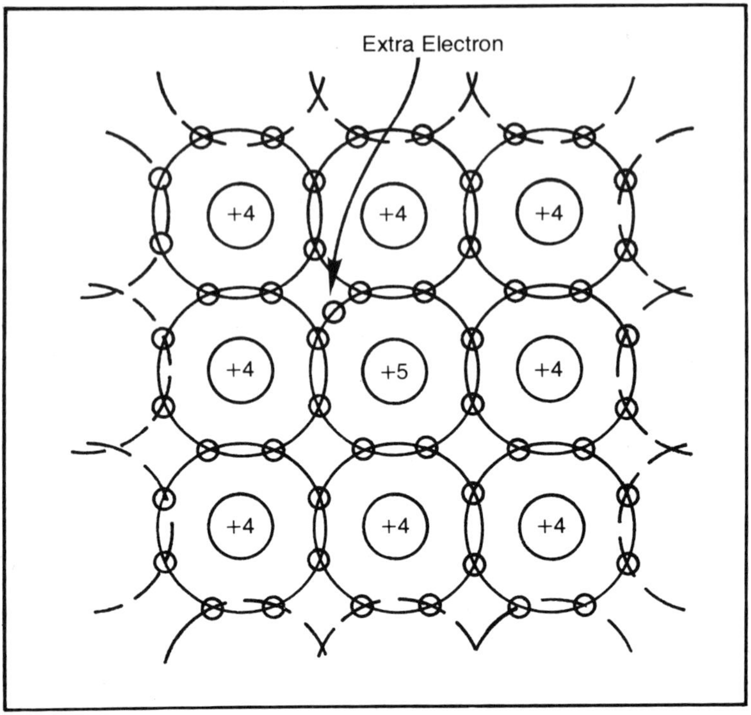

Fig. 2-7. Semiconductor atomic diagram with one atom of n-type material.

Looking at group 5b in Table 2-1, we see that phosphorus (P), arsenic (As), and antimony (Sb) all have the valence of 3. Therefore, they are the donors. These elements are also called *n-type* semiconductors, because they have one additional electron, or negative charge.

Now let us take a look at current flow in a semiconductor crystal with a small amount of n-type material. Figure 2-8 shows such a piece of semiconductor with three atoms of valence 3. If an electric field is applied across the crystal as shown, electrons will start to move. When the electron moves from the atom with an extra electron to the postive end, as shown by arrow 1, a positive charge is generated at the atom that the electron left. This is called a *hole*, or absence of an electron. The charge of this hole adds to the charge of the field to attract another electron. Since the extra electrons are the easiest to break loose from their parent atom, the electron from the next atom of n material will be attracted to the hole. This is shown by arrow 2.

This electron, leaving the parent atom, created another hole, and filled the previous hole. This hole, with the charge from the field, attracts the remaining extra electron, as shown by arrow 3. This positive hole attracts an electron from the negative side of the field, as shown by arrow 4. This completes the transition of one electron across the semiconductor. It is not important that it was a different electron that entered and then left the material, but only that an electron came out for every one that entered.

This process repeats millions of times during the act of conducting current through the semiconductor. The number of electrons flowing through the material per unit time determines the current flow. The amount of current depends on the value of the applied field and the amount of n-type material in the crystal. There are, in reality, several parallel paths of electron flow that exist at the same time. These paths may or may not interact with each other, and will change constantly.

Traditional current flow is from positive to negative, and is just a tradition. Electron flow is what makes up current flow, and is from negative to positive. This is the opposite to traditional current flow.

Study Fig. 2-8 and reread the above description of current flow in semiconductors until it is understood. The concept of replacing the extra electrons with other electrons must be understood to comprehend how semiconductors work.

Now let us add one atom of a material of valence 5 to a semiconductor crystal. This results in the absence, or shortage of one electron to fill the outer orbit and to be shared with an atom. This is as illustrated in Fig. 2-9. From the periodic table, atoms with the valence of 5, or three electrons in the outer orbit, are boron (B), aluminum (Al), gallium (Ga), and indium (In). Since this mate-

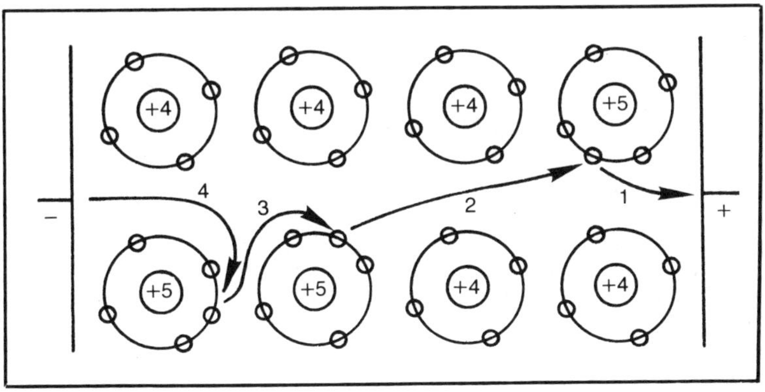

Fig. 2-8. Electron flow and movement in a n-type semiconductor.

rial has a shortage of one electron, it is called p-type material.

Referring to Fig. 2-9, the absence of an electron in the shared combination results in a positive charge (or minus electron) at that point. As mentioned before, this absence of an electron is called a hole. This hole can accept an electron to fulfill the requirements for each atom to share two electrons; therefore, it is called an *acceptor*, because it can accept an electron.

Current flow can be maintained across the semiconductor with a small amount of p-type material and with an applied field. Figure 2-10 shows this condition with three acceptor atoms and an applied voltage across the crystal. Thus, there are three holes within the structure. These holes, when assisted by the field, contain enough positive charge to capture electrons from other valence bands. Thus, in simplified form, the electron is conducted across the material by being attracted to the holes. When one hole is filled, another is generated.

Two types of conduction take place here. One is where the electron simply moves across the crystal by filling the first hole,

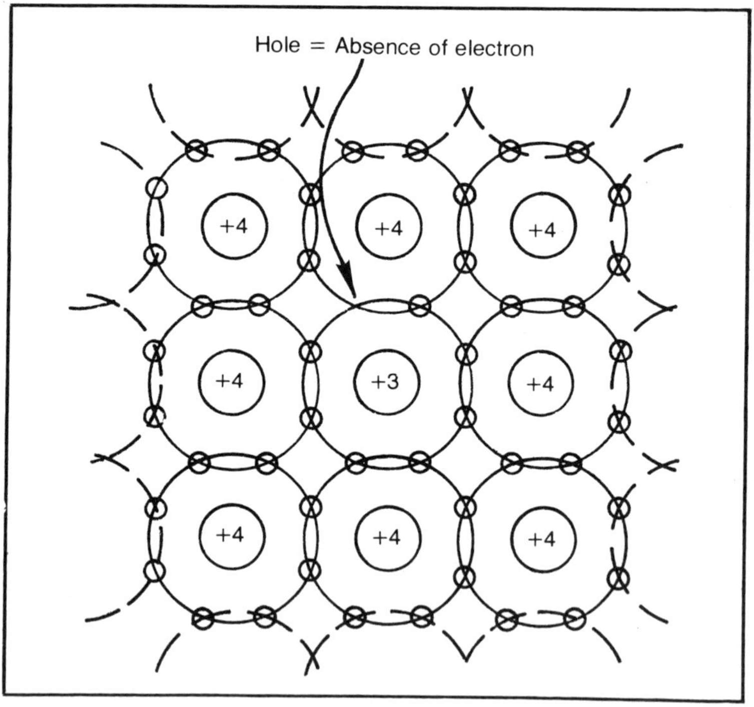

Fig. 2-9. Semiconductor atomic diagram with one atom of p-type material.

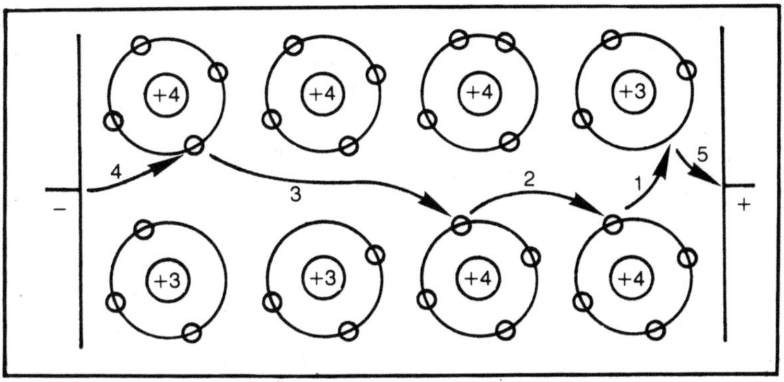

Fig. 2-10. The movement of an electron in a P-type semiconductor.

then being attracted to the second hole, and simply proceeding in this manner. The other type of electron flow is where a hole attracts an electron from an electron pair in another atom, as shown by arrow 1 of Fig. 2-10. Thus a hole is generated which attracts another electron from another atom (arrow 2). This newly generated hole attracts another electron from yet another atom (arrow 3). This hole is filled by an electron from the negative side of the voltage field (arrow 4). The first electron to move, now in one of the atoms of valence 5, moves to the positive side of the voltage field, thus completing the transition and allowing the next to start.

Although the electron moved from the negative side to the positive side, it can be conceived that a hole moved from the positive to the negative. This hole movement is in the reverse direction for arrows 1, 2, and 3. Thus, hole movement is the movement of the absence of an electron, and is in the reverse direction. This concept may be difficult to understand, but remember that hole movement is in reality the movement of a place in the valence orbit of different atoms where there is an absence of an electron.

An atom, which loses or gains one electron in the outer orbit is said to be *ionized*. A positive ion is an atom of the valence of 3 in which the extra electron has been removed and not replaced. The removing of the electron gives the atom a positive charge. Likewise, an atom with a valence of 5 that has an extra electron to fill the outer orbit has a negative charge. Therefore, it is a negative ion. The energy required to ionize an atom of a material is called the *ionization potential*. This is a definite value, and has different values for different elements. Ionization can go so far as to remove or add more than one electron to an atom, but this takes more energy. The energy required to ionize an atom completely is tremendous. Ionization, beyond the single charge level is a complex process and will not be dealt with here.

The pure semiconductor material, without any added atoms of valence other than 4, is said to be intrinsic. But, as we have seen, an intrinsic material is of little value for conducting current. For an intrinsic material to conduct current, it is required to break the bonds of electrons in the valence orbits which are shared by two atoms. This is a strong bond and is what holds the crystal together. The addition of an extra electron into the crystal structure by adding n-type material provides an electron that is not bound into the crystal structure. Therefore, the energy required to move this electron is considerably less. Likewise, the removal of one electron from the crystal structure by adding p-type material provides a positive charge. This positive charge, or hole, reduces the energy required to capture an electron.

When n- or p-type material is added to an intrinsic semiconductor, in reality, one atom of the semiconductor material is removed and replaced by the added material. This is done in the crystalline structure to obtain the desired results. This is called *doping* the semiconductor. The percentage of impurities (either n or p type material) is very low, and is determined by the final application for the semiconductor.

Figure 2-11 shows the materials used for the semiconductor and the materials used for doping the semiconductor.

THE pn JUNCTION

Now that we have looked at the individual p- and n-type semiconductors, let us explore what happens when we combine one p-type

Acceptors
- Boron (B)
- Aluminum (A1)
- Gallium (Ga)
- Indium (In)

Semiconductors
- Germanium (Ge)
- Silicon (Si)

Donors
- Phosphorus (P)
- Arsenic (As)
- Antimony (Sb)

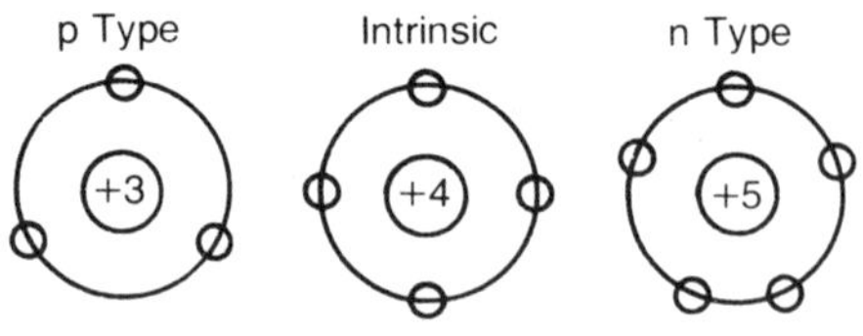

Fig. 2-11. Intrinsic, n- and p-type semiconductors.

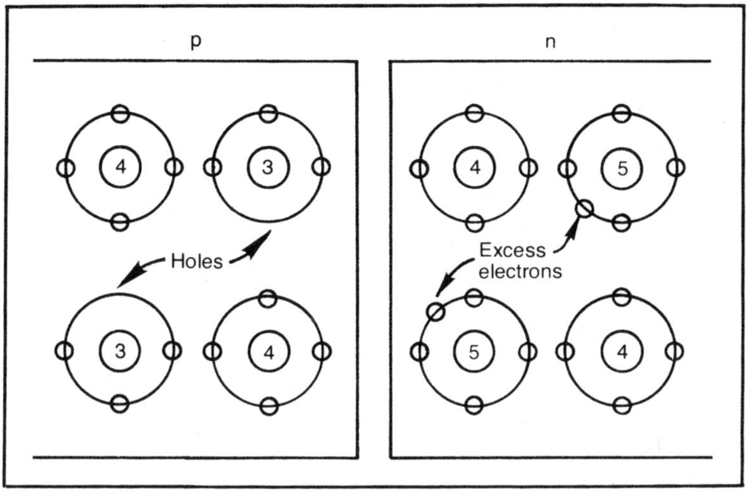

Fig. 2-12. A pn junction before forming.

semiconductor with one n-type, as shown in Fig. 2-12. This figure represents the junction just before forming. To form the junction, the two materials are made into one piece of semiconductor material with a p-region and a n-region.

The methods of actually forming the junction are many and may involve two pieces of semiconductor material. Or it may be one piece with a p-region and an n-region, both formed by the *diffusion* of the dopant material. Another method is the *deposition and etching* of the semiconductor. These methods will be covered in more detail later. For the present, it is sufficient to say that an atomic bond is formed between the two regions to make them into one crystal.

At the instant of forming the junction, the extra electrons leave the n-region and drift to fill the holes in the p-region, as shown in Fig. 2-13. This occurs only in close proximity to the junction. This is because as the electrons move, a positive region is set up on the n side of the juction and a negative region is set up on the p side. When the electrons leave the n-region, positive ions are formed. Likewise, when the electrons fill the holes in the p side, negative ions are formed. Remember that positive ions are atoms with one electron removed, and negative ions are atoms with one electron added.

It is these ions which form the positive and negative regions, as illustrated in Fig. 2-14. This figure shows the junction after it has been formed. The positive and negative regions extend outward

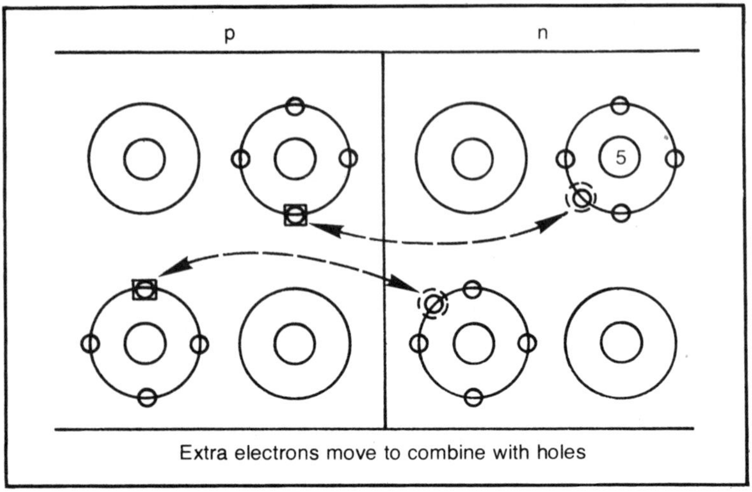

Fig. 2-13. A pn junction at the instant of forming showing electron and hole movement.

from the junction until the movement of electrons stops. This happens because the influence of the holes is limited by the distance and by the positive and negative regions. Figure 2-15 shows a more detailed view of the junction. Note that all electrons and holes have combined in the region of the junction to form what is called the *depletion region.*

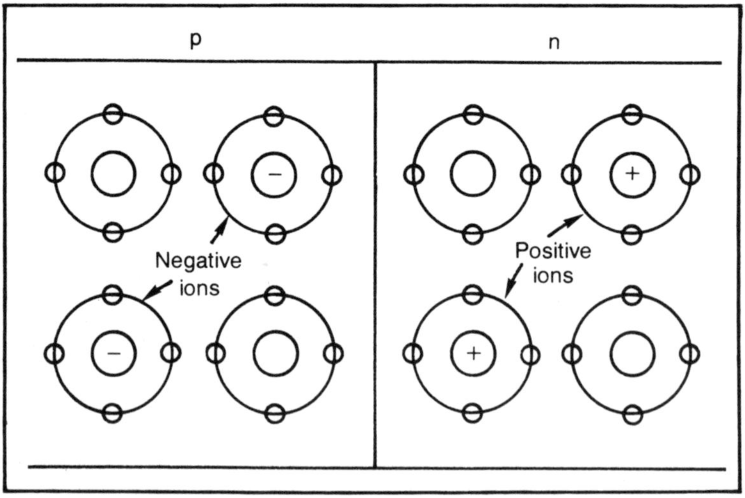

Fig. 2-14. A pn junction after it has stabilized.

The electric field is generated by the positive and negative ions formed in the depletion region with the positive side of the field in the n-region and the negative side in the p-region. This electric field is often referred to, and can be visualized as, a space charge *equivalent battery* with the polarity as shown. The field spans a specific distance, so it has width. It is of a given value so it has height. This value is the value of the equivalent battery and is 0.3 volts for germanium and 0.6 volts for silicon. This is the forward voltage drop of the pn junction.

Both the width and height of the field can be changed by applying an external voltage to the semiconductor crystal. De-

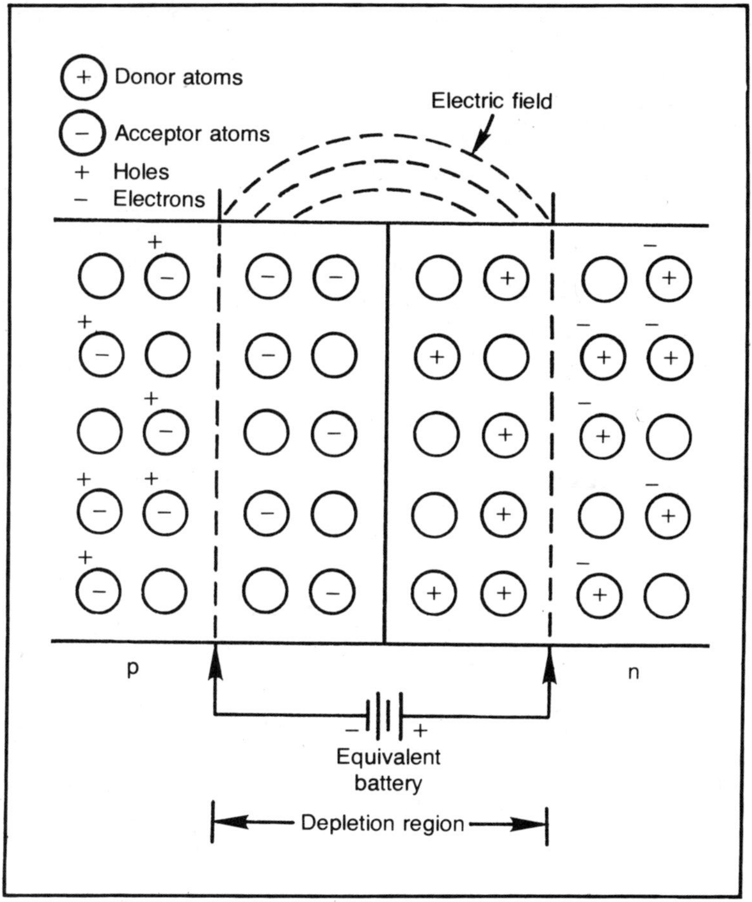

Fig. 2-15. Larger view of the pn junction showing the depletion region, electric field, and equivalent battery.

pending on the polarity and magnitude of this applied voltage, it will either assist, reduce or block current flow.

When a voltage is applied across the junction, it is either forward or reverse biased. If the positive voltage is applied to the n side of the junction, the junction is *reverse biased.* It is called reverse bias because it is in the opposite direction required to permit current flow across the junction. It adds to the equivalent battery. This condition is illustrated in Fig. 2-16.

When reverse bias is applied, more electrons flow into the p side, filling holes further from the junction. This increases the magnitude of the field and increases the width. For an electron to flow across the junction it must have enough energy to overcome this field, which takes considerable energy. If the reverse bias is increased above a certain point electrons will flow. This point is called the *reverse breakdown voltage*, which, if exceeded, can damage the junction.

For the reverse biased condition, the added voltage is of the same polarity as the equivalent battery. This increases the height and width of the field, and thus the depletion region. Therefore, it takes more energy to get an electron across the junction once the bias is established and the depletion region has stabilized.

If the external voltage is applied in the opposite direction, the junction is *forward biased.* For this condition, the negative voltage is applied to the n-region and the positive voltage is applied to the

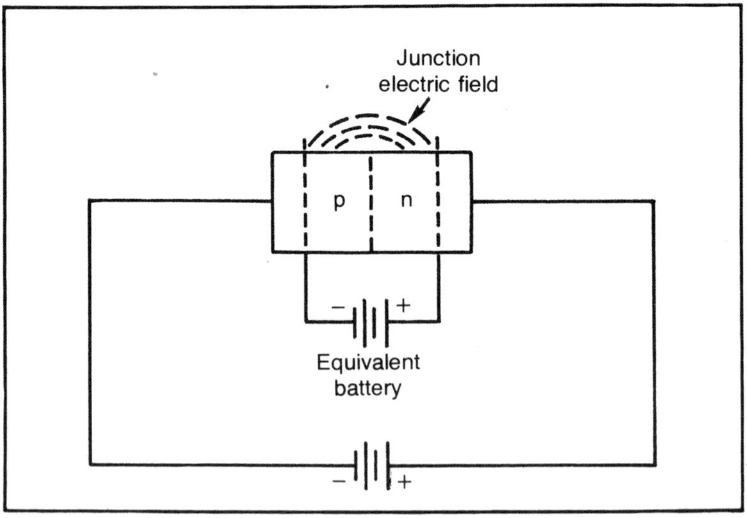

Fig. 2-16. A reverse biased pn junction showing the larger junction electric field.

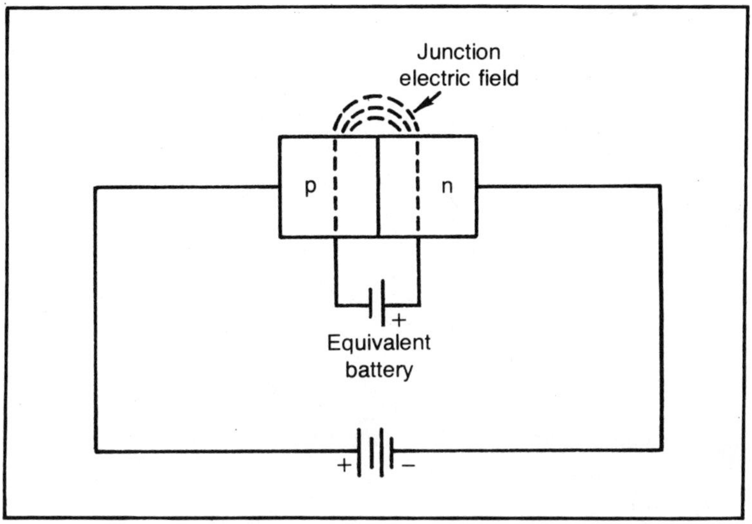

Fig. 2-17. A forward biased pn junction showing the reduced electric field.

p-region. This condition is illustrated in Fig. 2-17. Since the applied voltage is in the opposite direction from the equivalent battery, the net effect is to reduce the height and width of the junction electric field. There is a minimum value which the field can be reduced to, and increasing the voltage will not reduce it further. That value is the *forward voltage drop* for the semiconductor material used. These values are 0.6 volts for silicon and 0.3 volts for germanium and remain fairly independent of current flow.

For the forward biased condition, the junction electric field must be exceeded before current can flow. This is because the field forms a potential barrier that the electrons must overcome. Remember that electrons flow from the n-region to the p-region, while conventional, or traditional current flow is in the opposite direction. This pn junction can be used to form a junction diode. A diode conducts current in one direction and blocks it in the other direction. The positive side of the forward biased diode (or the p-region) is the anode, while the negative side (or the n-region) is the cathode. The convention and symbol for the diode is shown in Fig. 2-18. Some diodes are used in the reverse biased condition, so the anode is connected to the negative side of the source.

THE TRANSISTOR

The transistor is the active element in an IC, and as such, it must

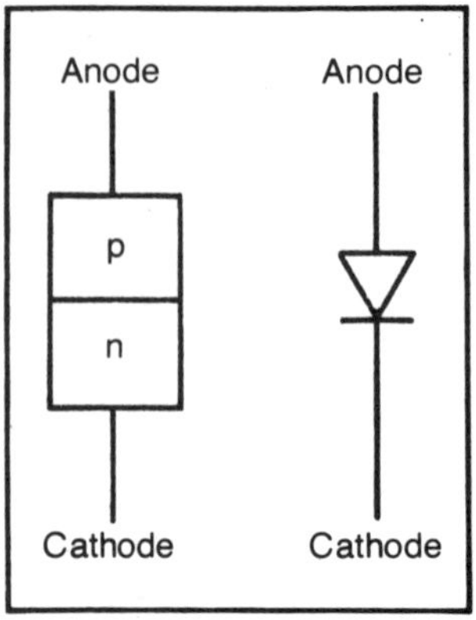

Fig. 2-18. Symbols for a pn junction diode.

provide useful gain for voltage and/or current. *Gain* is defined as the increase in amplitude from the input to the output. Voltage gain is the output voltage divided by the input voltage. Current gain is the output current divided by the input current. Power gain is the output power divided by the input power.

Before becoming involved in how a transistor works, let us look at an important concept. Figure 2-19 shows a hypothetical circuit, which if it would work, would provide power gain. In this circuit, R1 can be considered the input and R2 the output. With the switch in position A the battery is in series with R1. The current and power are calculated with Ohm's Law.

$$I = E/R1 = 10 \text{ volts}/10 \text{ ohms} = 1 \text{ Amp}$$
$$P = I^2\,R1 = 1^2 \times 10 = 10 \text{ Watts}$$

Now, if this same 1 amp could be made to flow through R2 when the

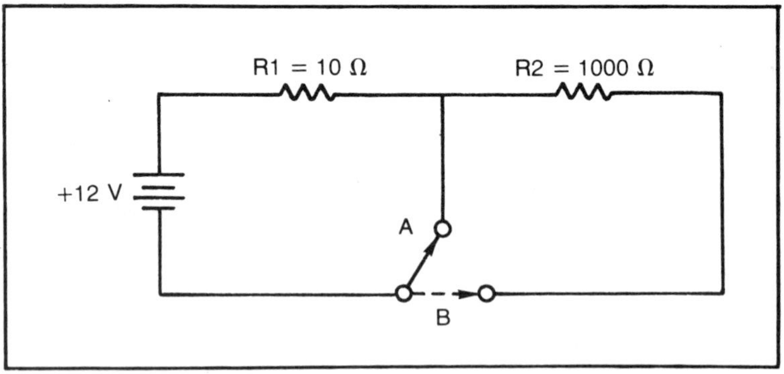

Fig. 2-19. Idealized resistor circuit with power gain.

switch is placed in position B, the power and power gain would be as follows:

$$P = I^2\ R2 = 1^2 \times 1000 \text{ ohms} = 1000 \text{ Watts}$$
$$\text{Power Gain} = P2/P1 = 1000/10 = 100$$

The power gain of this circuit is 100. 100 times as much power is at the output as we applied to the input.

This is an idealized illustration of how a transistor works. Since the circuit is fictitious and will not maintain the one amp of current flow through R2, other methods must be used to put this concept into practice. But this concept gives the transistor it's name. It's name is a combination of TRANSfer and reSISTOR.

Realize that gain is signal gain, and not the total applied power gain. The total applied power to an active device consists of a supply voltage or current and a signal voltage or current. It is the signal which is amplified or acted on to provide gain. The additional voltage or current comes from the supply voltage or current. If this was not the case, power could be generated by the circuit. The utilization of the supply voltage and current is the efficiency of the device.

To simulate the circuit shown in Fig. 2-19, let us combine a pn junction with an np junction so tht the n-region is common to both junctions. This forms a pnp transistor, as shown in Fig. 2-20. The

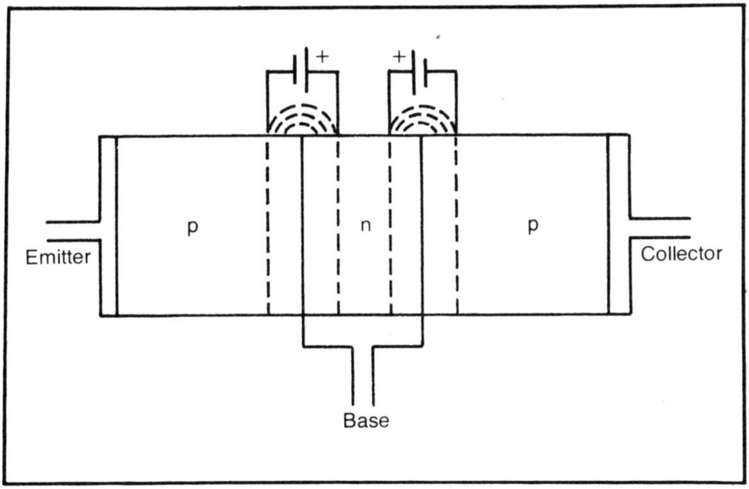

Fig. 2-20. A pnp junction transistor showing the two electric fields and equivalent batteries.

leads on a transistor are the *emitter, base,* and *collector*. They will be defined as to function in a later discussion. For, now, it is sufficient to say that they exist as shown. The width of the base is the width of the n-region. The equivalent batteries are shown at the top of the junction field, and the polarity is as shown with the positive sides closest together.

If the junctions are combined such that the p-region is common to both junctions, an npn transistor is formed as shown in Fig. 2-21. Notice that the equivalent batteries for a npn are in the opposite direction from the pnp transistor.

The base region is always the common region, while the emitter and collector are defined by the biasing of the junctions. The bias is applied from the emitter to the base and from the base to the collector.

By forward biasing one junction and reverse biasing the other junction, power gain is possible. The emitter-to-base junction is forward biased and the base-to-collector junction is reverse biased. This arrangement is shown in Fig. 2-22. The emitter injects *majority carriers,* either electrons or holes, into the base region. Those charges escaping recombination near the junction area will penetrate further into the base region. The amount of charges that escape recombination is dependent on the value of the applied bias. *Recombination* is the process of an electron and hole combining to eliminate each other.

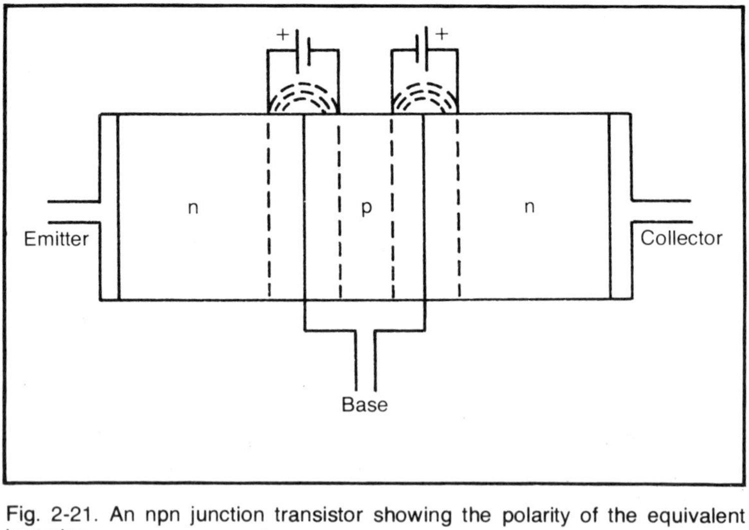

Fig. 2-21. An npn junction transistor showing the polarity of the equivalent batteries.

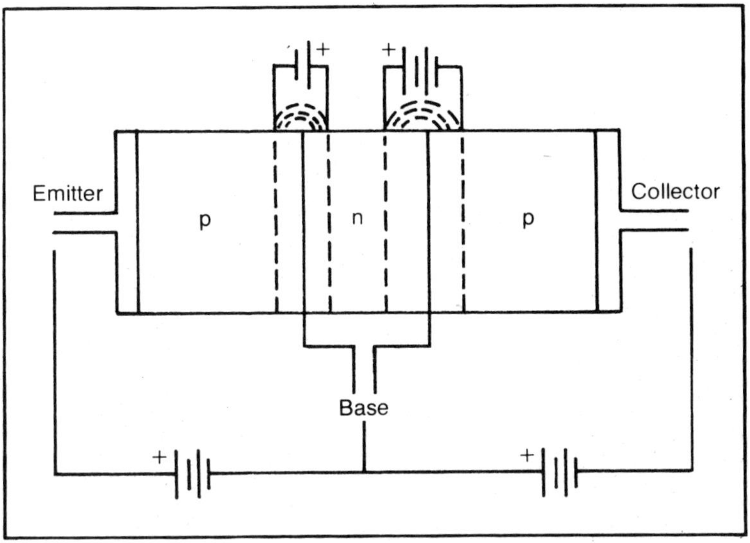

Fig. 2-22. A forward biased pnp transistor showing the decreased emitter-to-base junction electric field and the increased base-to-collector electric field.

Most of the charges escape recombination because of the reduced width and the height of the emitter-base junction electric field. It is reduced because of the forward bias applied to the junction. Those charges making it into the base region spread out by the process of diffusion. Applying reverse bias of sufficient magnitude to the base-collector junction will cause the charges to continue into the collector region, where they are collected. Majority carrier movement will therefore take place, originating in the low resistance region of the forward biased emitter-base junction, and continuing to the high resistance region of the reverse biased base-collector junction. This meets the requirements of the circuit shown in Fig. 2-19.

For the pnp transistor shown in Fig. 2-22, the *majority carriers are holes.* So holes are injected into the emitter. As they move from the emitter to the base, some combine with electrons to produce emitter-to-base current. Most of the holes injected into the emitter diffuse through the base and combine with electrons from the battery in the collector region. The high negative potential at the collector attracts these holes. Thus the collector current equals the emitter current minus the base current.

For the npn transistor shown in Fig. 2-23, the *majority carriers are electrons.* These electrons are injected into the emitter. As they move into the forward biased base region, some of the electrons

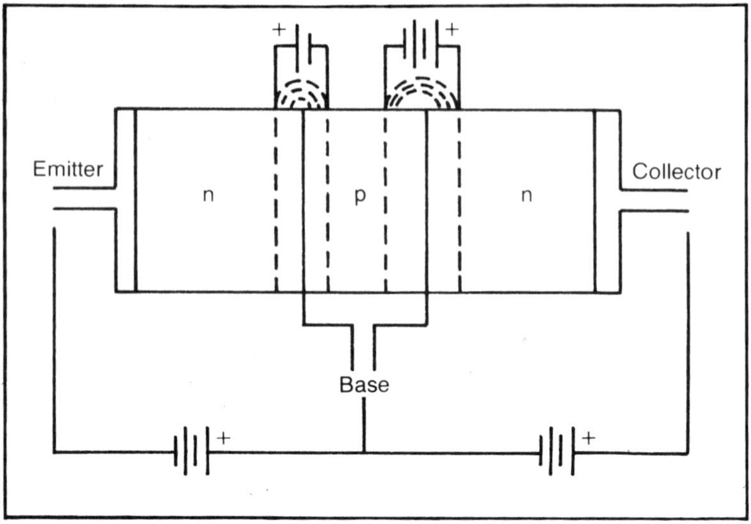

Fig. 2-23. A forward biased npn transistor.

combine with holes to produce emitter-to-base current. Most of the electrons diffuse through the base and combine with the holes at the collector. The high positive potential at the collector attracts the electrons.

A small change in the base-to-emitter bias voltage results in a large change in the number of electrons which cross the base region. This allows the base to control the emitter-to-collector current.

The transistors shown in Fig. 2-22 and 2-23 are connected in the common-base configuration. That is, the base is common to both the input and the output of the transistor. Figure 2-24 shows the common-base, common-emitter, and common-collector circuit configurations. Since the transistor is a three terminal device, one of the leads must be common to the input and the output. This fact is how the circuit configurations got their names. Table 2-4 gives some of the characteristics and advantages of each configuration. The configuration used depends on the application and the external circuitry. The common emitter is the most common because it has both voltage and current gain, but it also has phase inversion.

As mentioned before, there are two types of transistors, npn and pnp, depending on which material forms the base junction. Figure 2-24 shows npn transistors and the biasing required. Figure 2-25 shows both the npn and the pnp types and identifies the leads. The emitter always has the arrow, and the direction of the arrow

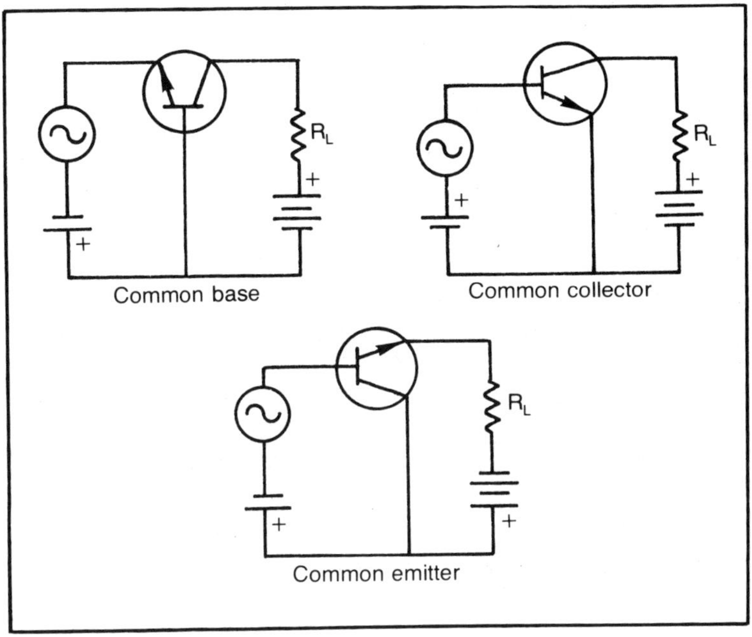

Fig. 2-24. The three circuit configurations for the transistor.

determines if it is an npn or pnp. If the arrow is pointing away from the base, it is an npn. If the arrow is pointing toward the base, it is a pnp.

Since the polarity of the biasing voltages for a pnp transistor is opposite that of an npn, if the transistors in Fig. 2-24 were replaced with pnp types, the batteries would have to be turned around. Also if there are any polarized capacitors or diodes in the circuit, they may need to be reversed.

Figure 2-26 shows a single stage RC-coupled transistor am-

Table 2-4. Characteristics of the Common Emitter, Common Base, and Common Collector Transistor Configurations

Characteristic	Common Base	Common Emitter	Common Collector
Power Gain	Yes	Highest	Yes
Voltage Gain	Yes	Yes	Less than 1
Current Gain	Less than 1	Yes	Yes
Input Impedance	Lowest	Intermediate	Highest
Output Impedance	Highest	Intermediate	Lowest
Phase Inversion	No	Yes	No

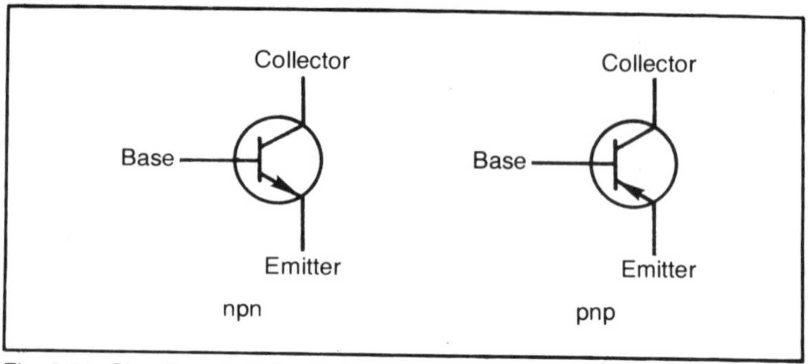

Fig. 2-25. Circuit symbols for the pnp and npn transistor.

plifier. R1 and R2 provide base biasing while R3 provides emitter biasing. R_L provides collector current with no input. The transistor shown is a pnp type, so the collector must be negative with respect to the emitter.

TRANSISTOR MANUFACTURING

There are many methods used to construct transistors. As the manufacturing technology increased and processes could be tightly controlled, more exotic and cheaper methods were developed. No matter what the method, they all start with an intrinsic semiconductor crystal of extremely high purity. The first step is to grow a

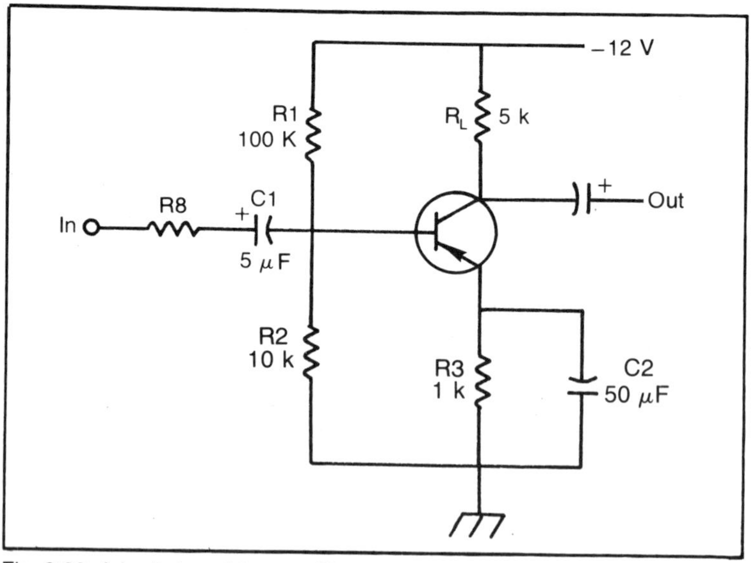

Fig. 2-26. A basic transistor amplifier circuit.

semiconductor crystal in a crystal growing device. A small seed crystal is slowly pulled from a molten bath of the semiconductor material under tightly controlled conditions in a controlled atmosphere. This yields one crystal typically about 6 inches long and 1 inch in diameter of good purity.

Since this crystal is not of high enough purity, it must be purified further. Higher purity is accomplished by a process called *zone refining*. It works on the principle that impurities will tend to stay in the molten material. Zone refining is the process of creating a zone of molten material that slowly moves down the length of the crystal. The material on both sides of the zone is solidified, so the crystalline structure is maintained.

To achieve a crystal of high enough purity (less than 5 parts impurity per 100 million parts of semiconductor) may require more than one pass through the zone refiner. When the refining is completed, the impurities will be confined to the last part of the crystal to pass through the refiner.

This part of the crystal is sliced off, and the impurities are discarded. Then the crystal is sliced into thin wafers to be used in making transistors and ICs.

To make junction transistors, these smaller pieces are remelted, and smaller crystals are regrown. This time, as the crystal is grown, some donor element is added to the molten bath, and the semiconductor becomes an n-type semiconductor. Later in the process enough acceptor material is added to the molten bath to overcome the effect of the initial n material and to provide the required doping level for the p-region. The semiconductor that is now solidifying is p-type. Thus, at the transition, a pn junction is formed. The area around the junction can be sliced out to make junction diodes.

If n material is now added to the molten material, another junction is formed, thus forming an npn region. This region is sliced out, cut up and packaged to form npn transistors. Several transistors are made from one of the remelted crystal slices.

To form pnp transistors the same process is used, only p material is added first, then n material, then p material again.

This process has several variations, dealing mainly in the method used to apply the dopant. One method is to draw the remelted crystals in alternating n and p atmospheres. Thus several pnp or npn regions can be grown in the same crystal. Another variation is to use a zone refining type of process to apply the dopant to the crystal.

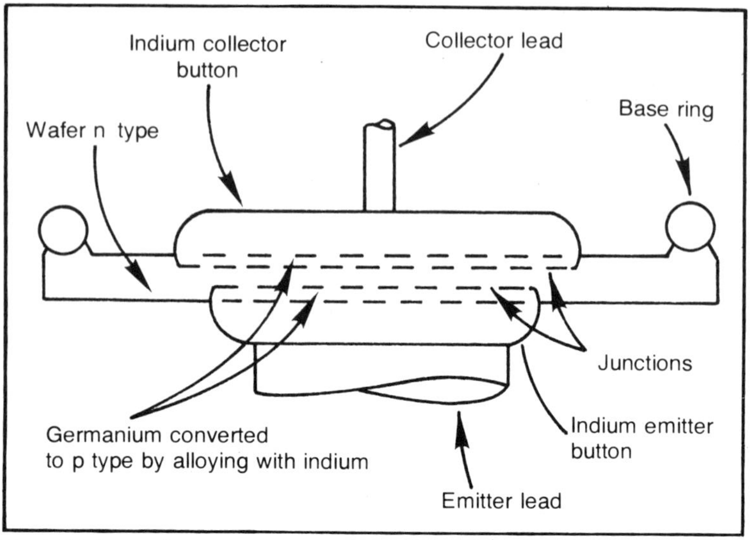

Fig. 2-27. Internal construction of an alloy power transistor.

Another method used in making transistors is the *alloying process*, as illustrated in Fig. 2-27. In this transistor, the semiconductor wafer is doped with a n type material. Then two buttons of a p type material are alloyed into opposite sides of the wafer. Indium is used because it has low resistance and it will alloy with germanium. The wafer, with the indium buttons with leads attached, is temperature cycled to the alloying temperature. Then the indium buttons wet and alloy into opposite faces of the wafer, forming the junctions. Thus a pnp transistor is formed. The two indium buttons become the emitter and collector, and the original n type wafer is the base.

Another method of making transistors is by *deposition and diffusion*. This method is described in the next chapter.

3

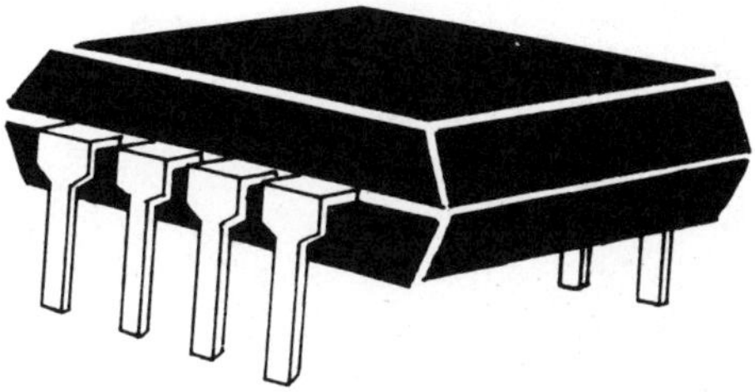

Integrated Circuit Manufacturing

The early integrated circuits were simple devices in which the transistor and resistor version of a digital logic gate was constructed as an integrated circuit. Such a device is shown in Fig. 3-1. This is the schematic for a dual gate. This IC consists of four transistors and six resistors. It was originally developed as a digital logic circuit, but it soon found many other applications, including use as an analog amplifier. This and other similar integrated circuits gave the electronics engineers, designers, and experimenters an integrated circuit to work with. Before long they started asking the semiconductor industry for more and different ICs. Several manufacturers started making ICs of different types and circuit designs. The new ICs were all digital integrated circuits in the early days because these were the easiest to build and the component tolerances were not as stringent as for analog circuits.

Where transistors resulted in a tremendous space savings over tubes, integrated circuits resulted in a considerable space savings over transistors. Figure 3-2 shows some transistorized printed circuit boards and the equivalent integrated circuits. The cost savings in the printed circuit boards due to the reduced size was a significant amount, and in some cases, it was more than the cost of the integrated circuits.

Using integrated circuits, the circuitry that was on several printed circuit boards was now able to be mounted on one printed circuit board. This allowed complex functions to be accomplished

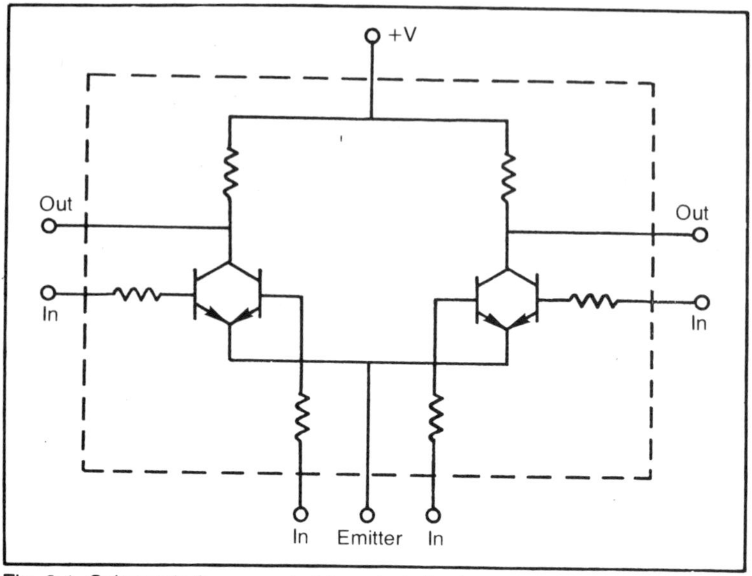

Fig. 3-1. Schematic for an early integrated circuit.

economically and with a minimum of parts. But this was just the beginning. As the years progressed and the technology improved, more and more components were packed into an IC. Soon it became

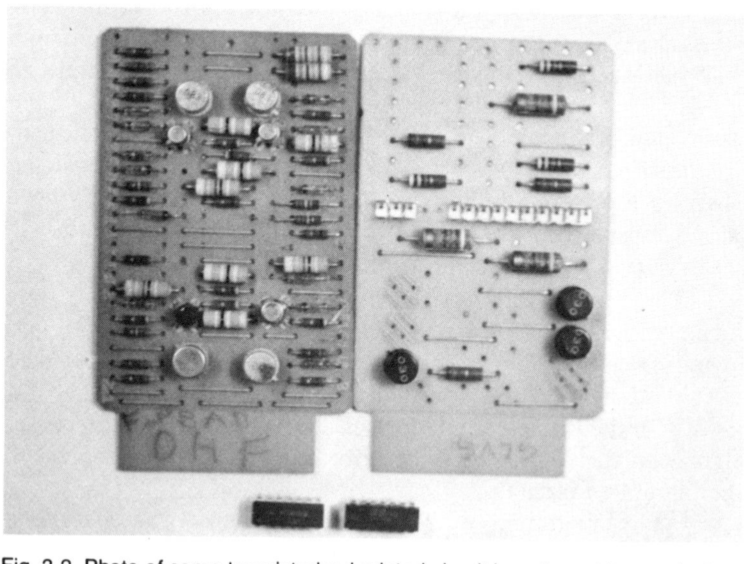
Fig. 3-2. Photo of some transistorized printed circuit boards and the equivalent integrated circuits.

a problem of the physical size of the leads used in the integrated circuits themselves. These were restricting the amount of components which could be packed into an area.

THE PROCESS

In the previous chapter, the theory of a pn junction and the operation of transistors were covered. Integrated circuits use junction transistors usually made up on a substrate of silicon. Several different manufacturing processes are used, and the main steps are masking, diffusion, deposition, and etching.

Masking is the application of a mask material to the chip to prevent the next process from effecting the masked area. *Diffusion* is the process of one material diffusing into the atomic structure of another so that the atoms of the added material replace atoms of the original material in the crystal lattice. Only certain materials will diffuse with other materials, and it usually requires heat to initiate the process.

Deposition is the depositing on one material onto another. This forms a layer of the deposited material on the original material and may or may not form an atomic bond. Deposition is usually accomplished by vaporizing the material to be deposited with heat or some form of bombardment. The vaporized material is made to strike the item it is to be deposited upon, either by electrical field, normal convection, or by saturating the atmosphere and letting the vaporized material condense. It is usually done in a controlled atmosphere to prevent impurities from contaminating the material.

Etching is the removal of unwanted material by some process. This can be chemical etching or electric etching. Normally, chemical etching is used. The material to be removed is dissolved away by a chemical. Typically, the area not to be etched is protected by a mask which is resistant to the etchant.

In the early integrated circuits, there were three different types of devices on the chip: transistors, resistors, and conductors. The conductors connected the various transistors and resistors to each other and to the pads which led to the IC pins.

Conductors are made by depositing a metallic conductor on the chip. Since the length of the conductor is short, there is very little effect caused by the resistance of the conductor. Resistors are made by either using a resistance material deposited on the chip or by using a controlled pn junction.

Integrated circuits are built up of successive layers of material on a substrate. The collector and emitter regions are deposited on top of the substrate. Over these are the conductors and the resistors. Insulating material is used to maintain isolation.

Growing Transistors

Let me explain the process of making a hypothetical transistor for an integrated circuit. The process is illustrated in Fig. 3-3, which shows the cross section of the chip as the transistor is being assembled.

Step 1 shows the silicon chip, polished and lapped until it is smooth. Next (step 2) a mask is placed on the chip to outline the area of the transistor. Then the pure silicon is diffused with some p-type impurity to change the area not masked into a p-type semiconductor. Then the mask is removed, and a layer of n-material is deposited over the total area (step 5). Then, in step 6, the area for the emitter and collector is masked. The n material not protected by the mask is etched away in step 7. This leaves the emitter and collector masked, and with n material under the mask. In step 8, a conductive layer is laid down, and this layer masked for the base contact and the remaining etched away. Then an insulating layer is deposited. Next the mask is removed, exposing the emitter, collector, and base contacts, all separated by an insulating material. The remaining steps, not shown, are to mask for the conductors to the base, emitter, and collector contacts, and to deposit the conductors.

This represents only one very crude method of making a transistor for the IC. While the transistor is being assembled, other things are happening on different areas of the chip. Each of the steps covers the total chip, so many transistors can be grown at the same time.

From the above example, it becomes apparent that the key to the fabrication of the device is the mask. This is what allows the different processes to effect the various areas of the chip. The mask is placed on the chip by a photographic technique. The material is covered with a photosensitive material and then exposed to a photographic negative of the mask. This chemically changes the photosensitive material of the exposed area so that it will not rinse away. In this way, the pattern of the negative is reproduced on the chip. The clear areas of the negative are not rinsed away and become the mask.

Each of the masking steps becomes one step in the artwork for the integrated circuit. This artwork is made, either by hand, by computer, or manually with machine assist. Since the finished artwork is intricate in detail and the dimensions are extremely

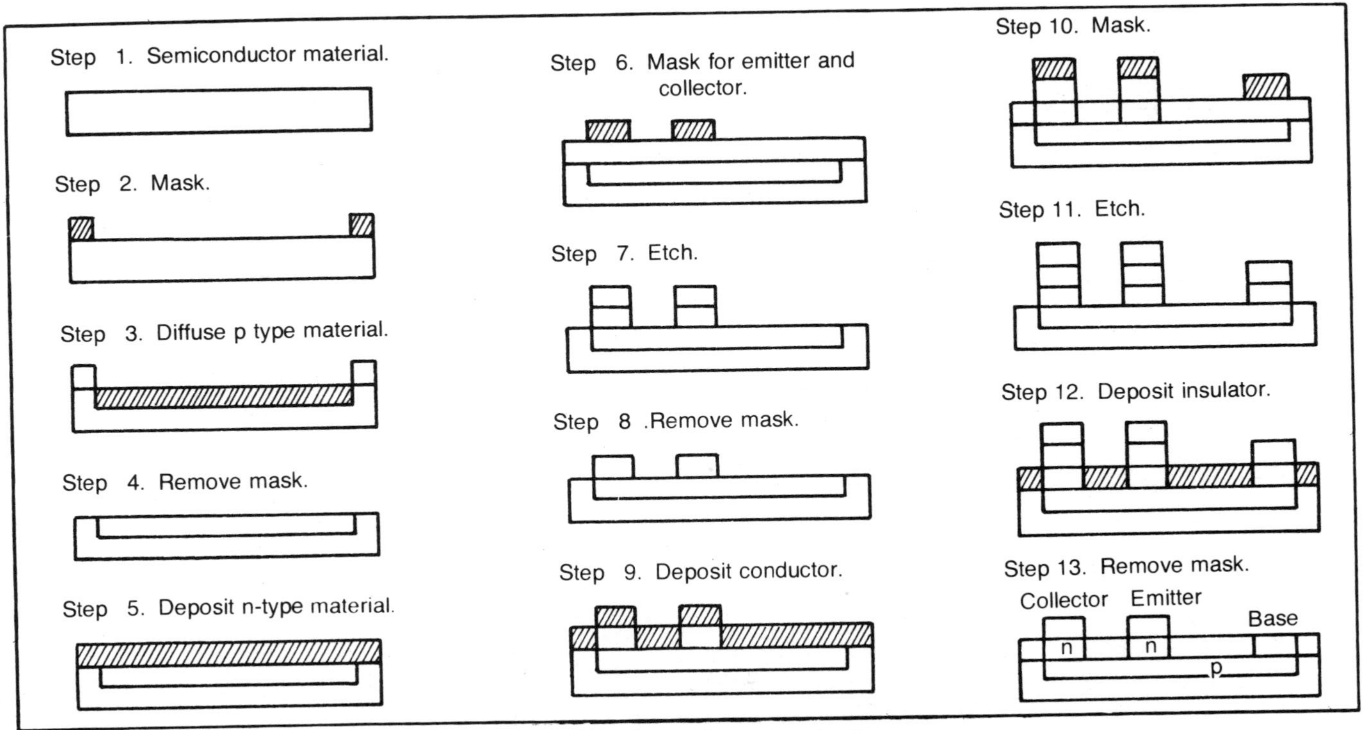

Fig. 3-3. Illustrating the steps in the fabrication of part of a hypothetical integrated circuit.

small, the artwork is made several times real size. This makes it easier to work with, and the tolerances are not as tight. When the artwork is complete, it is photographically reduced and several of the images are placed on one tool film. The tool film is what is used to expose the photosensitive layer on the integrated circuit. It is the same size as the silicon wafer which is being processed. Depending on the chip size and complexity, each wafer may contain from less than a hundred to over a thousand individual chips.

After the wafer has been processed to grow or make the integrated circuit, each individual chip is tested by automatic test equipment. Due to the small size of the IC chip, special probes are used to make contact with the circuit. Those chips which fail the testing are marked for removal later. When all the chips on the wafer have been tested, the wafer is cut into the individual chips by a diamond cutter, either a scribe or a saw. Then the failed chips are discarded and the good chips are mounted on the lower half of the package.

Bonding

Depending on the chip size and the configuration of the pads to provide the electrical connection to the outside world, two different methods of bonding are used. One method is to bond the chips, right side up, to the package base. Then small gold wires are bonded to the chip pads and to the package pads leading to the pins, thus providing the connections to the pins on the IC package.

The second method is to turn the chip upside down and bond it to the pads on the package. Either a low melt solder or a conductive adhesive is used to bond the chip pads to the pads on the package. This second method lends itself to automatic assembly, but requires a package land pattern for the chip and special equipment. This method is often called *flip bonding* and cannot be used if the top of the chip must be exposed for external processing, as in some of the new computer integrated circuits like PROMs.

LIMITATIONS

There are some factors which limit the amount of components that can be placed on an integrated circuit chip.

☐ The resolution that can be obtained for the individual lines on the chip.

☐ The separation which must be maintained between the lines on the chip.

☐ The insulation required between the individual components on the chip.
☐ The surface area available on the chip.
☐ The yield of good chips.

The first two items are determined by the resolution of the optical system used in reducing the IC layout drawing to the chip size. If the resolution is not good enough, the lines become fuzzy and the control of the size and boundary is lost. This has been improved considerably by using better and more advanced optical systems to

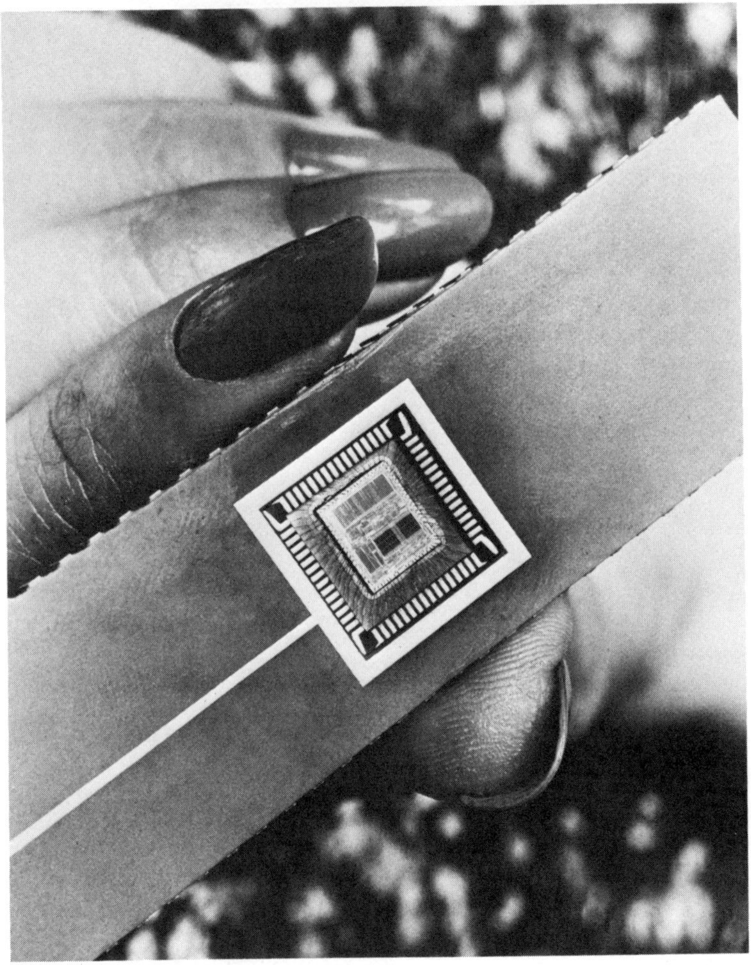

Fig. 3-4. Enlarged view of a Motorola MC68000 integrated circuit (courtesy of Motorola Semiconductor Products, Inc.).

Fig. 3-5. Enlarged view of the MC68000 IC chip after fabrication and before cutting the wafer (courtesy of Motorola Semiconductor Products Inc.).

the point where the actual conductor width is becoming a limiting factor.

The insulation required between the individual components is improving by using reversed biased junctions and better insulation material. The surface area available has increased as the handling techniques have been refined. The problem here is the chip substrate, which is fragile and easily broken.

The *yield* is the percentage of good chips obtained from a manufacturing lot. The yield is directly dependent on the number of

Fig. 3-6. Two IC chips directly mounted to the pads for the package leads.

components on the chip. The more components and conductors present, the greater the chances for a failed unit. As the design and processes improved and new technology was incorporated, the yield improved dramatically. When an integrated circuit is first introduced, the yield may be low. But, as more units are produced and the failure modes are identified and corrected, the yield improves.

Present day technology has improved to such a state that

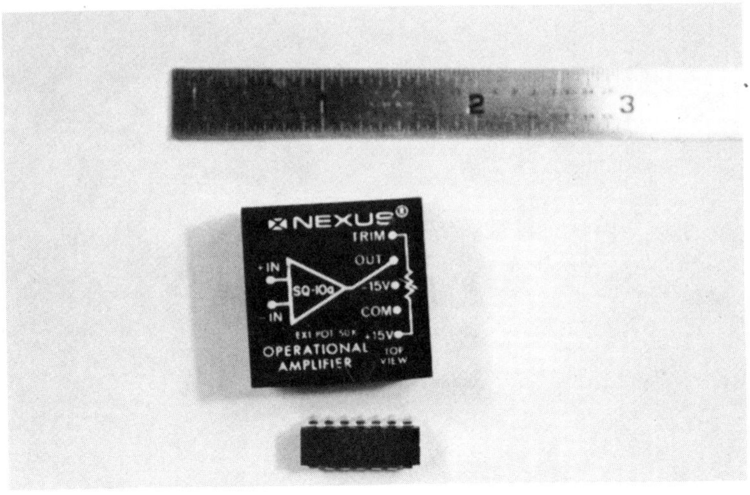

Fig. 3-7. A size comparison of a hybrid operational amplifier and an integrated circuit operational amplifier.

microprocessors for small computers have been made into a single chip. Figures 3-4 and 3-5 show one such unit, the Motorola MC68000. The device shown in Fig. 3-4 is the packaged unit which is about 3 ¼ inches long and about ⅞ inch wide. This is a 16-bit microprocessor housed in a 64-pin package. Figure 3-5 shows an enlarged view of the chip itself. This chip is 246 mils by 281 mils (.246 by .281 inches). It contains at least 68,000 transistors. This photograph shows the chip as part of the wafer. The black line completely around the chip marks the perimeter of the chip. Additional chip boundaries are visible on all four sides of the chip.

The external pads are around the edge of the chip. In Fig. 3-4, the interconnection of these pads with the pads on the package can be seen, and consist of small gold wires.

Figure 3-6 shows a chip mounted upside down on the package, with the pads on the chip bonded directly to the pads on the package. These chips are about 120 mils square.

HYBRID CIRCUITS

Parallel development in the analog circuit area consisted of developing hybrid circuits. These circuits consisted of combining transistor chips and resistors and capacitors, either as chips or as discrete components, onto a substrate. For some of the applications, these were only used until the integrated circuit equivalents

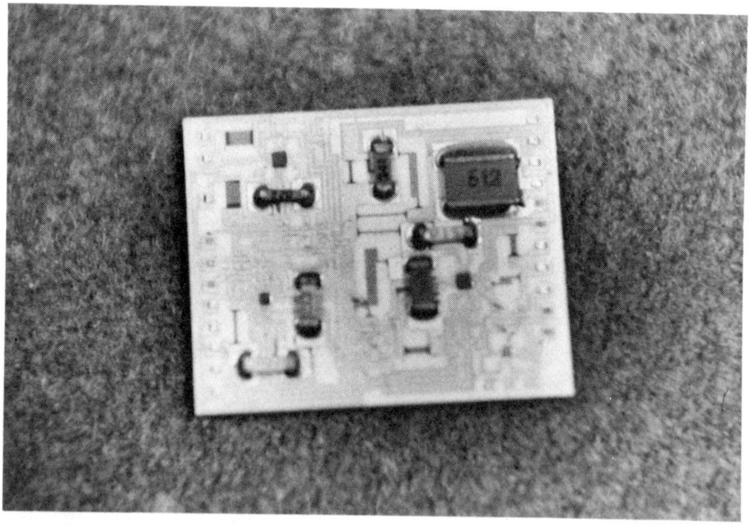

Fig. 3-8. Hybrid circuit showing chip resistors and IC chips mounted directly to pads on the substrate of the package.

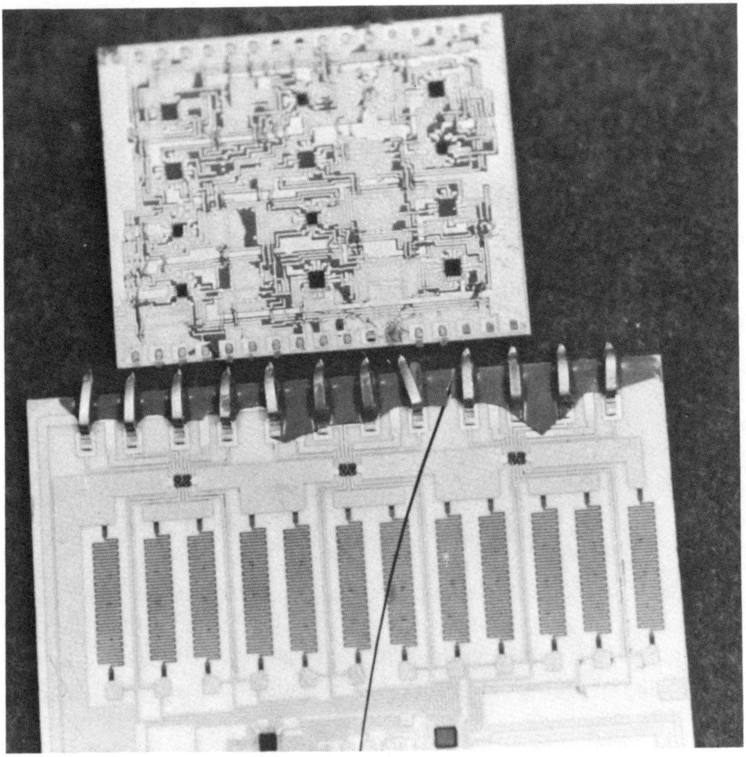

Fig. 3-9. Two custom packaged integrated circuits.

were developed. Figure 3-7 shows a hybrid operational amplifier, and an integrated circuit operational amplifier. The IC device is much cheaper because it is manufactured with automatic equipment.

Figure 3-8 shows another hybrid device, which uses some chip components and chips attached to a substrate to make the circuit. This type of assembly requires a minimum of manual labor, and can use assembly equipment.

Another development in the process of achieving larger and more complex chips is to assemble several smaller chips onto one substrate and assemble them into a circuit. Figure 3-9 shows one such chip. It contains 12 smaller chips connected to achieve a complex function. The lower chip of that figure shows a hybrid with 5 chips and 12 resistors. This example shows what can and is being done in the advancing field of integrated circuit technology.

4

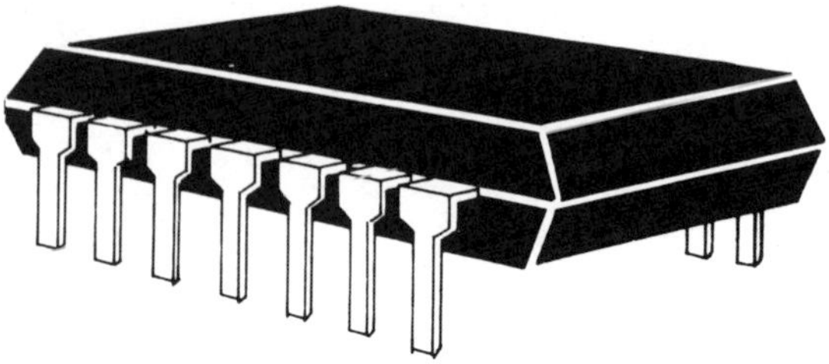

Digital and Analog ICs

There are hundreds of different types of integrated circuits, covering a multitude of functions. These range from simple logic gates to complex microcomputers and control integrated circuits. These complex chips may have thousands of elements on one chip. The first major division as to the types of ICs has to do with the circuit contained in the IC. This broad division is analog and digital. This chapter introduces you to both types of circuits and describes the operation of some of the different types of functions.

Digital circuits operate with the requirement that they have only two stable states, full on and full off. *Analog* circuits operate with changing voltage levels between the full on and full off condition. For some circuits, it is not readily apparent if it is an analog circuit or a digital circuit, while other circuits leave no doubt. Some ICs may fall into both categories. Analog circuits are sometimes used in digital applications and digital circuits are sometimes used in analog applications.

DIGITAL

Digital circuits are circuits which have two stable states, full on and full off. The states in between are not allowed by circuit design. The exact amplitude of the output voltage is not of prime importance. It is the state of the output that is important. These states are called by many names.

Full on/Full off
True/False
1/0
On/Off

These are called the logical states of the circuit, and can be equated to the simple circuit shown in Fig. 4-1. For this circuit, the light is on when the switch is closed and off when the switch is open. So, when the switch is open it represents the off or false state. When the switch is closed it represents the true, or on state.

Logic functions are the functions that digital circuits perform. They are the operations on the input states that can be done to achieve the desired output. The terms *digital circuits* and *logic circuits* are used interchangeably and have the same meaning. Logic functions will become better understood as we proceed.

There are three basic logic functions: AND, OR, and NOT. The *logic equation* is a mathematical expression for how a digital circuit operates in the input states to produce the output. The overall circuit may be one individual circuit or a combination of logic circuits, each with their own logic equations. Mathematical symbols are used to represent the logic functions. A dot between two characters represent an AND function, a + represents the OR function, and an overscore over a character represents the NOT or invert function. Alphabetical characters are used to represent the inputs and outputs. Simple logic equations might look like these:

$$A \cdot B = C$$
$$A + B = C$$
$$\overline{A} = C$$

The first equation reads, input states A **and** B must be high to generate a high at output C. Otherwise, output C will be low. Logic equations imply that if the conditions of the equation are not met,

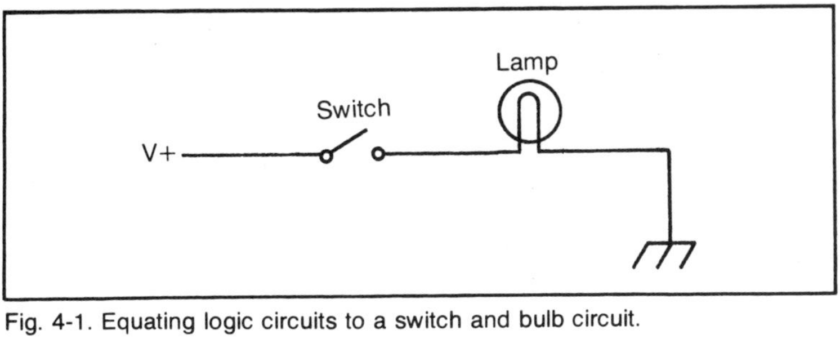

Fig. 4-1. Equating logic circuits to a switch and bulb circuit.

the output will be in the opposite state. For the top equation, if either A or B is low, the output will be low.

The second equation reads, input states high for either A **or** B result in output C high. Otherwise output C will be low. The last equation reads, output C is the **inverse** of input A. Sometimes this is read as output C is the NOT function of input A. Both are correct.

The logic equation $A + B = \overline{C}$ is read: the NOT function of output C is when inputs A or B are high. The NOT function is a low so the equation also reads: inputs A or B high results in output C being low.

Truth tables are used quite often in the analysis of logic circuits. They are also very handy in describing the logical operation of a circuit. The truth table is a table which lists all the inputs, and gives the outputs for all possible combinations of inputs. It is set up with one column for each input on the left and one column for each output on the right. Usually the first line is for all inputs equal to zero, and a line dedicated for each different combination of inputs. The outputs for each line is determined from the logic equation or the logic circuit. Table 4-1 is the truth table for the logic equation $A + B = C$, which reads, either A high or B high results in output C being high. Columns A and B are for inputs, and column C is for the output. The first line of the table is for A and B low (zero). When A and B are low, the output C is low. The remaining three lines show the conditions for when one or both of the inputs are high. This results in a high output.

If the logic circuit has more than one output, and they are not interrelated, it may require one truth table for each output. If the outputs are related, they can be shown on the same truth table. Table 4-2 shows the truth table for two equations: $(A \cdot B) + (\overline{C} \cdot D) = E$, and $E + \overline{F} = G$. This equation represents a logic circuit with two outputs, and the second is dependent on the first output. Notice that output E is just to the right of the inputs which generate that output. Input F is next.

Parenthesis are used in the equation to group functions together. Without the parenthesis, the equation reads differently

A	B	C
0	0	0
0	1	1
1	0	1
1	1	1

Table 4-1. Truth Table for the Basic OR Gate

A	B	C	D	E	F	G
0	0	0	0	0	0	0
1	0	0	0	0	0	0
0	1	0	0	0	0	0
1	1	0	0	1	0	1
0	0	1	0	0	0	0
1	0	1	0	0	0	0
0	1	1	0	0	0	0
1	1	1	0	1	0	1
0	0	0	1	1	0	1
1	0	0	1	1	0	1
0	1	0	1	1	0	1
1	1	0	1	1	0	1
0	0	1	1	0	0	0
1	0	1	1	0	0	0
0	1	1	1	0	0	0
1	1	1	1	1	0	1
0	0	0	0	0	1	1
0	0	0	1	0	1	1
0	1	0	1	0	1	1
0	1	1	0	0	1	1
1	0	0	0	0	1	1

Table 4-2. Combination Truth Table for Two Equations

because the functions are done in order, instead of those inside the parenthesis first. Study the equations and derive the truth table. It will help understand how they are made.

Basic Logic Functions

As stated before, there are three basic digital functions, AND, OR, and NOT. The remaining functions are derived from these. There are some complex functions, but only the basic three will be discussed here. Some of the complex functions are briefly described elsewhere in the book.

OR. The first function to be analyzed is the OR function. That is, in simplified form, the circuit represented by the equation $A + B = C$. Which means that either **A or B** high results in C high. Figure 4-2 shows the switch and light equivalent of a two input OR gate. If either switch A or switch B is closed, the light is on. The truth table for this circuit is given in Table 4-1. Follow through each line of the table by closing the switch to represent a 1, and opening it to represent a 0. For the output, a 1 represents the light on and a 0 represents the light off. Notice that the only time the light is off is when both switches are open.

AND. The next function to be covered is the AND gate. The

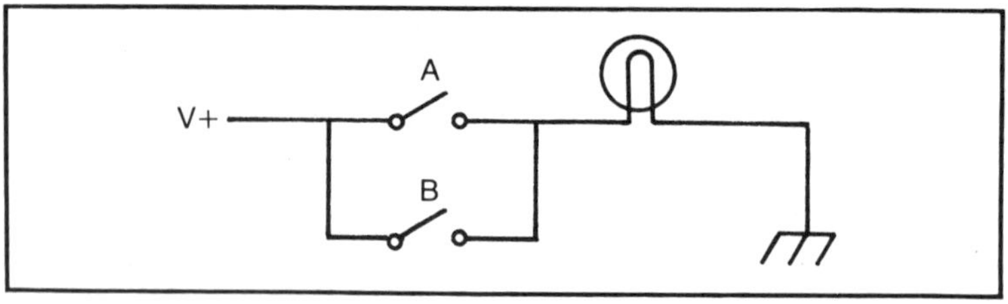

Fig. 4-2. Switch and bulb OR circuit.

simplified logic equation for this gate is A • B = C. Inputs **A and B** high result in output C high. Figure 4-3 shows the switch equivalent for such a gate, along with the truth table. Repeat the exercise of closing each switch and determining the output for all combinations. Notice that the only time the light is on is when both switches are closed.

Two input gates are used here to illustrate the AND and OR functions, but the gate may have more inputs. A three input gate will have eight possible combinations of inputs, but the same rules apply. An OR gate output will be high when one or more of the inputs are high. An AND gate output will be high when all the inputs are high. When in doubt about the state of the output, make up a truth table.

NOT. The NOT function is generated by an inverter. This gate provides the NOT function of the input, and the logic equation $\overline{A}$ = B. The truth table is given by Table 4-3. This gate is used to change a signal to its opposite state.

A *driver,* sometimes called a buffer, closely resembles the inverter. It is a single stage noninverting circuit. The logic equation is simply A = B, and the truth table is given in Table 4-3.

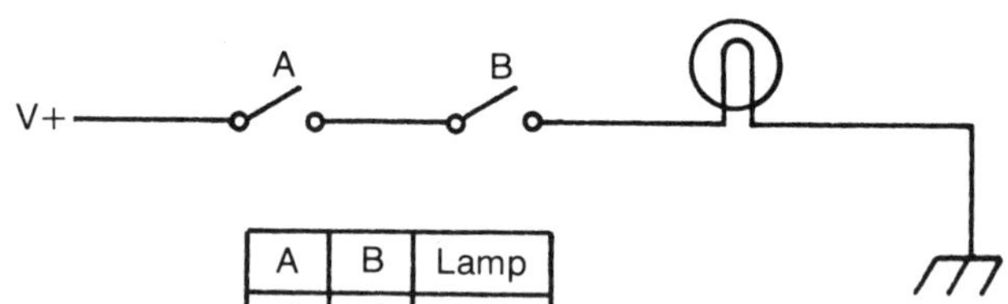

A	B	Lamp
O	O	Off
O	C	Off
C	O	Off
C	C	On

O - Open
C - Closed

Fig. 4-3. Switch and bulb AND circuit.

Inverter

A	B
0	1
1	0

Driver

A	B
0	0
1	1

Table 4-3. Truth Tables for an Inverter and Driver

NOR. There are two remaining common gates which are derived from the AND and OR gates. These are the NAND and NOR gates. These are simply the basic gates with a NOT output. They can be conceived as the AND and OR gates with an inverter added. The NOR gate is a **not** OR gate, and the logic equation is $A + B = \overline{C}$. This reads, A high **or** B high results in output C low. The truth table is shown in Table 4-4.

NAND. Similarly, the NAND gate is a **not** AND gate. The logic equation is $A \bullet B = \overline{C}$. This means, A and B are high to produce C low. The truth table is shown in Table 4-5.

Table 4-4. Truth Table for an AND Gate

A	B	C
0	0	1
0	1	0
1	0	0
1	1	0

Study these gates carefully and understand the logic function of each, for they are the building blocks for most digital circuits. Combining these gates to achieve more complex functions is covered in a later chapter. The operation of the gates is illustrated in Chapter 8.

Figure 4-4 shows the logic diagrams, logic equations and the circuit symbols for each of the above gates. The small circle on the output of the inverter, NOR gate, and NAND gate signify that the output is active low. That is, the output is low when the input conditions for the logic equation are satisfied.

A	B	C
0	0	1
0	1	1
1	0	1
1	1	0

Table 4-5. Truth Table for a NAND Gate

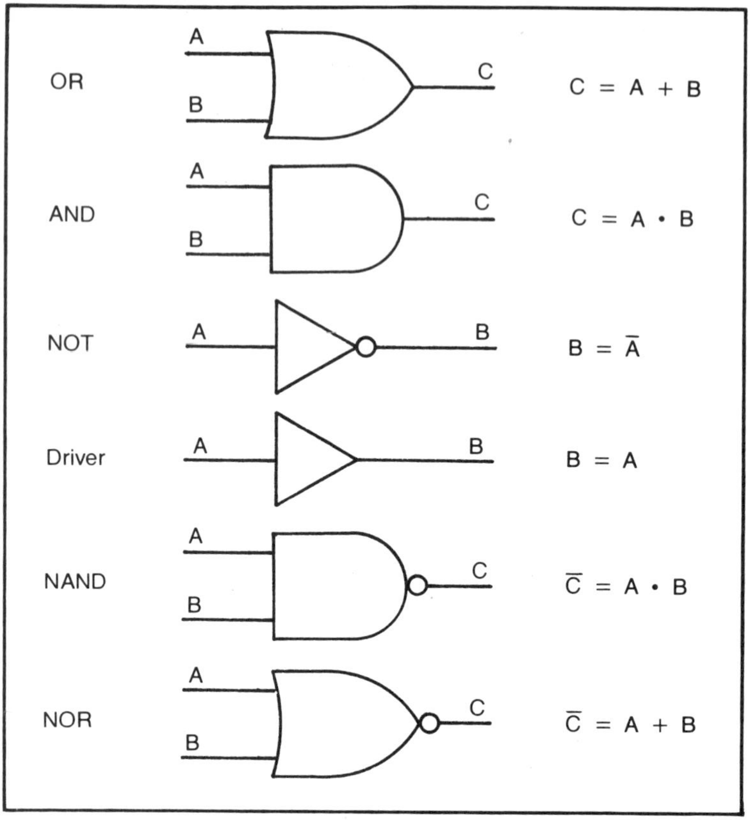

Fig. 4-4. Circuit symbols for basic logic gates.

Transistor Switching

In digital circuits, the transistors are usually configured as common emitter, and set up as switching circuits. Figure 4-5 shows a basic transistor switching circuit. A high input turns the transistor on. If the transistor is on, it can be considered as a short, so the output will be low. If the transistor is off, it can be considered as an open. So the output will be a high. When the output is high, current flows through the collector resistor to the load. Any current drawn will increase the voltage drop across the resistor, reducing the voltage amplitude at the output.

Transistor switching circuits depend on the ability of the circuit to switch from the full on to the full off state, and not to achieve any of the states between the full on and full off. Figure 4-6 illustrates this. This figure shows a load line drawn on a set of transistor

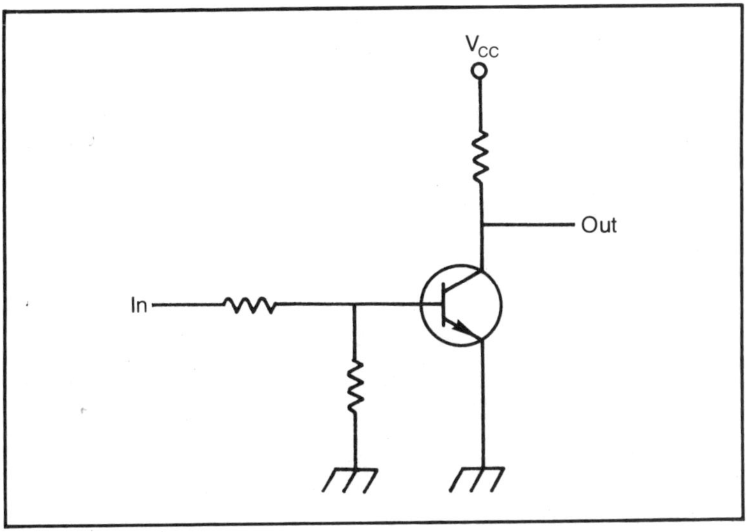

Fig. 4-5. Simple transistorized switching circuit.

output curves. In the idealized state the full off state is with no collector current, and the full on state is with no collector to emitter voltage drop. This condition is approached by driving the transistor from full off to saturation, for a small change in the input voltage. The input bias establishes the voltage level when this occurs.

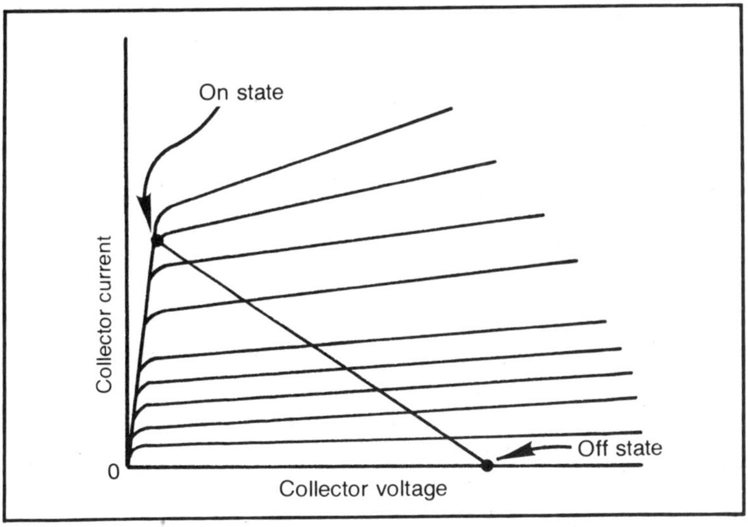

Fig. 4-6. Transistor response curve and load line for a switching circuit.

The standard common emitter switching circuit provides inversion, that is, when the input is high, the output is low.

The *transition time* between the two states takes the transistor into the region of high power dissipation momentarily. The time for the transition is determined by the response time of the transistor. If this transition time is excessive, the power dissipation can damage the transistor from overheating. Therefore, care must be taken to use input signals with a fast rise and fall time, and not to load the signals to prevent full voltage transitions for high and low.

Digital Families

There are several different families of digital ICs. These families represent different circuit components, various active elements, and different approaches to switching circuits.

RTL	Resistor Transistor Logic
DTL	Diode Transistor Logic
TTL	Transistor Transistor Logic
CMOS	Complementary Metal Oxide

The first two families were part of the development of the current digital families, and are not used at the present time. The third, TTL, is the most popular digital family for simple logic circuits. It is this family that is responsible for the wide usage of logic ICs. They are made in such great quantities that the cost is extremely low. Low cost means that these ICs can be used at a savings over discrete components.

CMOS circuits are ICs which use field effect transistors as their active element. This FET transistor has some advantages over the standard transistor, such as high impedance and low current requirements. But they also have disadvantages such as noise susceptibility and the ability of being damaged by static electricity. Despite these problem CMOS, or some offshoot of it, is the family of the future.

RTL. The resistor transistor logic family uses resistor logic circuits as the logic elements driving a transistor. Resistors are used to bias the transistor so that it is in the off condition (open) when there is no input, or the input is low. When the input is high, the transistor turns on, driving the output low. Figure 4-7 shows the basic RTL NOR gate. The truth table is also given. Notice that this is almost identical to Table 4-4, with another input added.

When all the inputs are low, the base of the transistor will be at

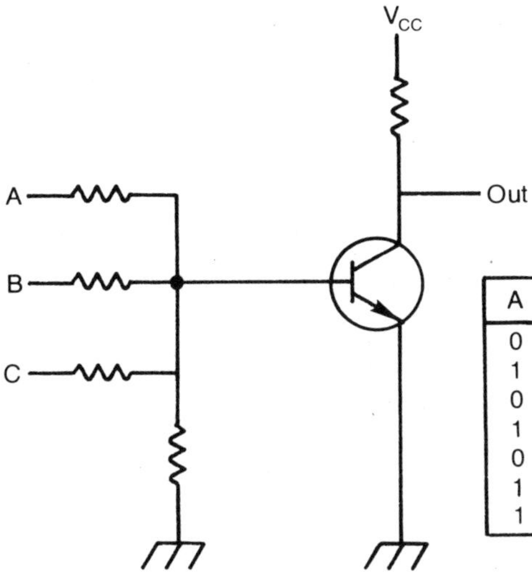

A	B	C	Out
0	0	0	1
1	0	0	0
0	1	0	0
1	1	0	0
0	0	1	0
1	0	1	0
1	1	1	0

Fig. 4-7. Basic RTL NOR gate with the truth table.

ground. Since there is no emitter base bias in this condition, the transistor is off. If the transistor is off, there is no collector-to-emitter current flow, therefore the collector voltage will approach VCC. Any current drawn by the output will decrease the value of the collector voltage, and thus the output voltage, by an amount equal to the output current times the collector resistor. Since the output is high for a low input, this circuit inverts the input.

When one of the inputs is high, there is forward bias generated across the base-emitter junction. Thus the transistor will conduct current. If the resistor values have been chosen correctly, the transistor will turn full on. If it does not turn full on, the transistor may not reach the on state as shown in the response curve of Fig. 4-6. If this happens, the transistor may overheat and burn up. When the transistor is on, there is collector to emitter current. This results in a low output. If two of the inputs were high, the operating point will be almost the same. The transistor will be driven slightly more into saturation.

Therefore, if one or more of the inputs are high, the output will be low. This satisfies the requirements for a NOR gate.

To make this NOR gate into an OR gate requires the addition of an inverter, as shown in Fig. 4-8. The inverter is a single input NOR gate, operating in the same manner as described above. When the input is high, the transistor is on, resulting in a low output.

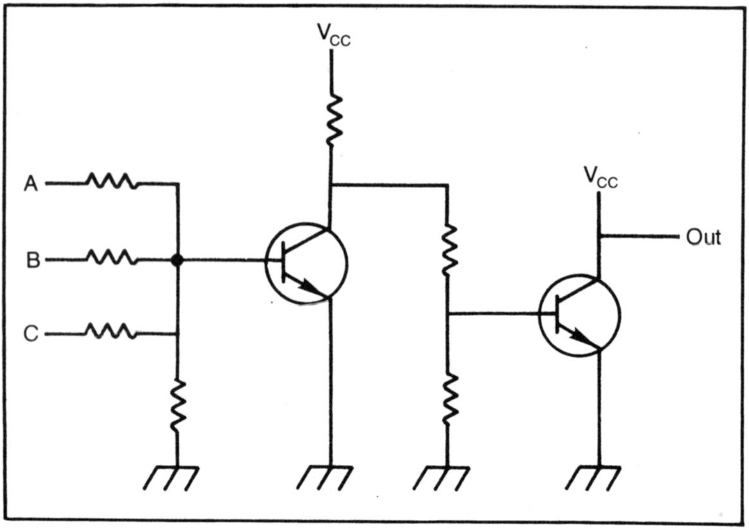

Fig. 4-8. RTL OR gate.

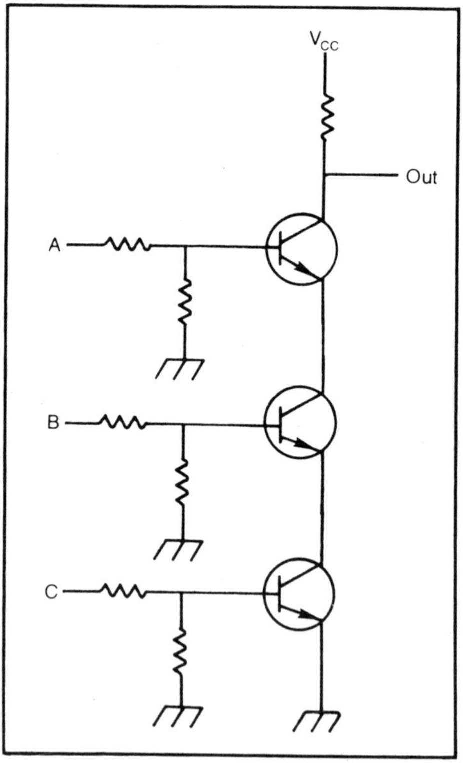

Fig. 4-9. RTL NAND gate.

Figure 4-9 shows one of the circuits used for the RTL NAND gate. For this circuit, all the transistors must be conducting before any collector to emitter current can flow. Therefore, all the inputs must be high to cause all the transistors to conduct current, resulting in a low output. If any one of the inputs is low, that transistor will be off, therefore no collector current will flow. This results in a high output. The output voltage for the low output condition will be the sum of the collector to emitter voltages when the transistors are on. This can result in the low output voltage being a volt or more. To overcome this, most NAND gates of this type have two inverters following the gate. This provides a low voltage of less than ½ volt for the on condition.

Figure 4-10 shows the circuit for the RTL AND gate. This is,

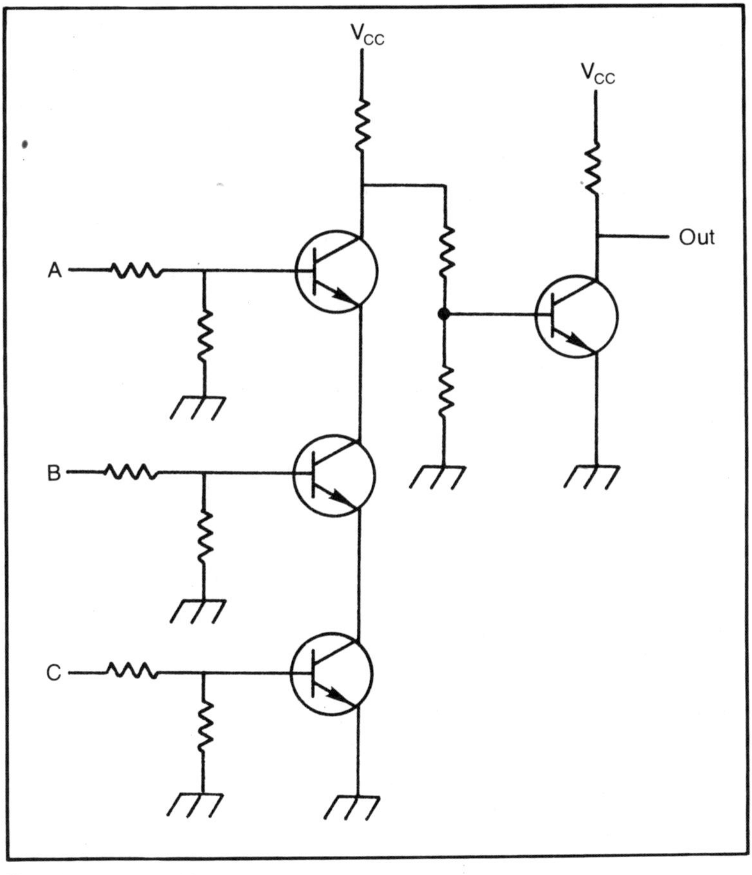

Fig. 4-10. RTL AND gate.

simply, the above NAND gate, with an inverter added. There are other AND gate circuits, but on some of them the bias is difficult.

Although there were many circuits used for RTL logic, these few serve to illustrate their operation. RTL was used during the infancy of ICs. Most of the RTL ICs were simply the discrete component circuits committed to an IC. These circuits were some of the logic circuits that were in use before ICs were constructed.

As the speed and complexity of logic circuits increased the problems with RTL logic limited its use. Some of the problems were switching speed, partial switching, drive current, and bias current. There were some circuit modifications which overcame some of the problems, but the main problem of speed remained.

DTL. The diode transistor logic family is the next step on the evolution of logic circuits. As with the RTL family, discrete DTL logic circuits were in use before the advent of the logic ICs. As the name implies, DTL circuits use diodes for the logic, and the transistor as the active element. The active element is required to maintain the logic levels. Without this active element, the low voltage level will increase and the high voltage level will decrease as more gates process the signal.

Figure 4-11 shows the basic DTL NOR gate. If all the inputs are low, there is no current flow through the diodes. Therefore, the base is at the ground level, and there is no base-emitter bias. No current will flow. This condition results in a high output, if all inputs are low.

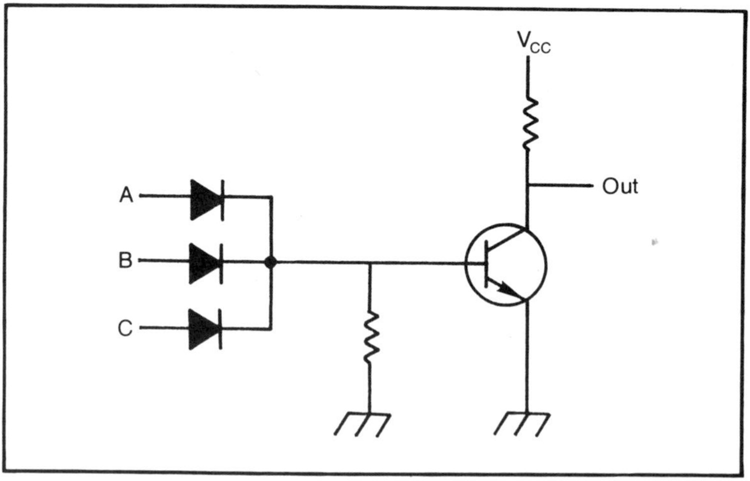

Fig. 4-11. DTL NOR gate.

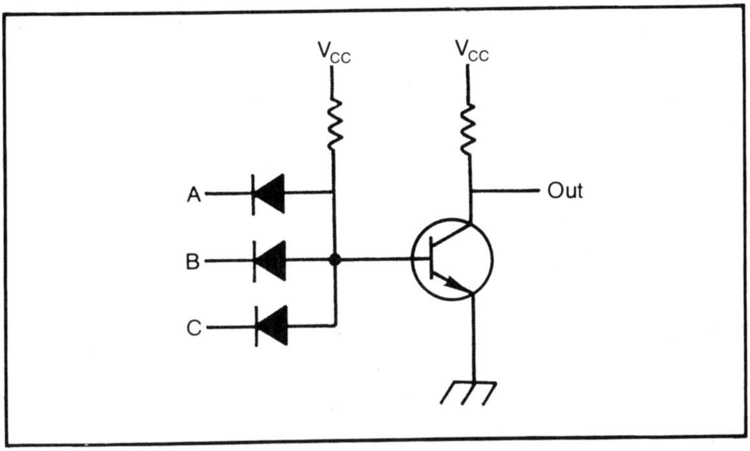

Fig. 4-12. DTL NAND gate.

If any one of the inputs are high, then current will flow through that diode. Since all the other diodes will be back biased, they will not conduct current. The one diode conducting current, will supply base current for the transistor, and bias current through the bias transistor. Thus, the transistor will turn on. This results in a low output. The conditions for a NOR gate are met. Any high input results in a low output. The basic NAND gate is as shown in Fig. 4-12. This circuit was also a discrete component logic circuit made from resistors, diodes and transistors. It was converted to an IC.

For this NAND gate, if all the inputs are high, the base voltage will approach the applied voltage. There will be some current through the bias resistor for base current of the transistor. There will be no diode current. The base emitter junction will be forward biased, therefore the transistor will be on. This condition results in a low output for all the inputs high.

If any one of the inputs is low, that diode will conduct current. This will clamp the base of the transistor to ground, turning the transistor off. This happens because the base-emitter junction is no longer forward biased. If the transistor is off, the output is high. These conditions meet the requirements of the NAND gate, that is, if all the inputs are high, the output is low. If one or more of the inputs are low, the output is high.

For both the NOR and NAND gates, adding an inverter to the output converts them to OR and AND gates respectively.

Figure 4-13 shows a more advanced DTL NAND gate. This uses several transistors to enhance the switching speed and im-

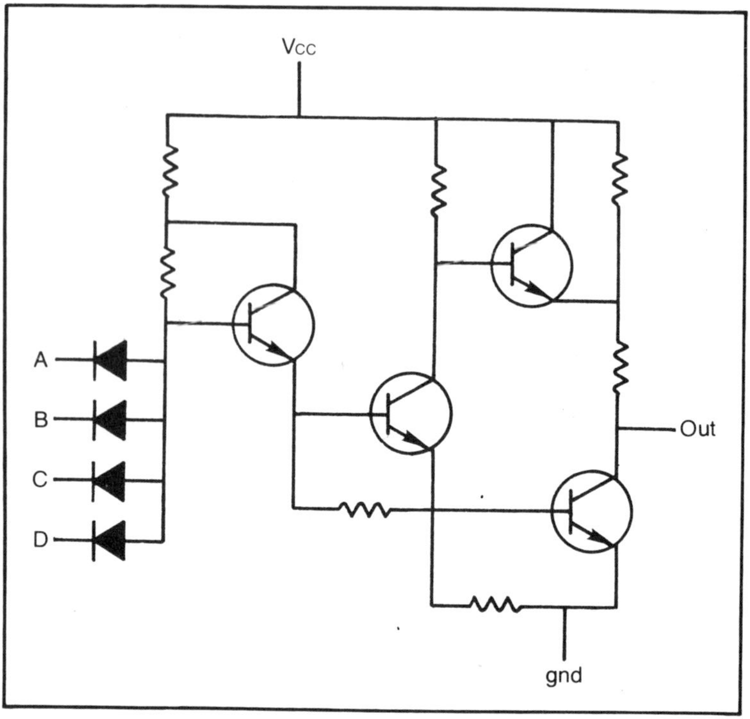

Fig. 4-13. Advanced DTL NAND gate.

prove the current capabilities. But there are still problems with the DTL circuits, which limited their use.

TTL. The transistor-transistor logic family uses transistors as the logic elements and as the active elements. This family was designed from the ground up for integrated circuits, and has no working discrete component equivalent. Although circuit diagrams are given for these gates, these are the IC circuit diagrams. TTL gates use multiple emitters as the logic elements, and these are constructed on the IC substrate.

The most popular of the TTL logic ICs are the 7400 series. This series contains a hundred or so different types of gates, with at least four versions. The versions each offer an advantage over the others, and each version has its uses. The 7400 series is popular because it is versatile, cheap, and available from several manufacturers. For quite a while it has been the standard of the industry, with clear cut and established specifications.

The three popular versions, besides the standard 7400 series are:

☐ H type—Gold doped structure with higher power and faster speeds.
☐ S Type—Uses the Schottky clamped transistor technique to make very high speed, high performance devices.
☐ LS Type—Uses low power Schottky devices combined with clamped transistors to make a low power high speed device.

The differences between the four versions are given in Table 4-6. Sometimes the devices can be used interchangeably, and other times they cannot. Usually a S or LS type device can be used in place of a standard device of the same number. The 7400 series are numbered with a 74 as the first two numbers. Next is the version, which is a character for any version except the standard. Then the gate number, which is two or more numbers. Any letters which follow are modifiers, such as military, plastic, ceramic, etc.

For example, 7402 represents one type of gate of the standard version. The H type version is 74H02, the S version is 74S02, and the LS type is 74LS02. The function of all these gates is the same but some of the characteristics are different.

In this book, we will discuss and use the standard 7400 series digital ICs. This is because these are the most inexpensive, and easy to obtain.

Figure 4-14 gives the circuit for a two input NAND gate. Transistor Q1 is the logic transistor, and the number of emitters determines the number of inputs for the gate. This transistor is set

Table 4-6. Comparison of the Standard, H, S, and LS Series 7400 Type TTL ICs

	Standard	H Series	S Series	LS Series	Units
High Level Output Voltage	2.4	2.4	2.5	2.7	
Low Level Output Voltage	0.4	0.4	0.5	0.5	
High Level Input Current	40	50	50	20	μ
Low Level Input Current	−1.6	−2	−2	−.36	mA
Low Level Output Current	16	20	20	8	mA
Gate Delay	10	6	3	9	nsec

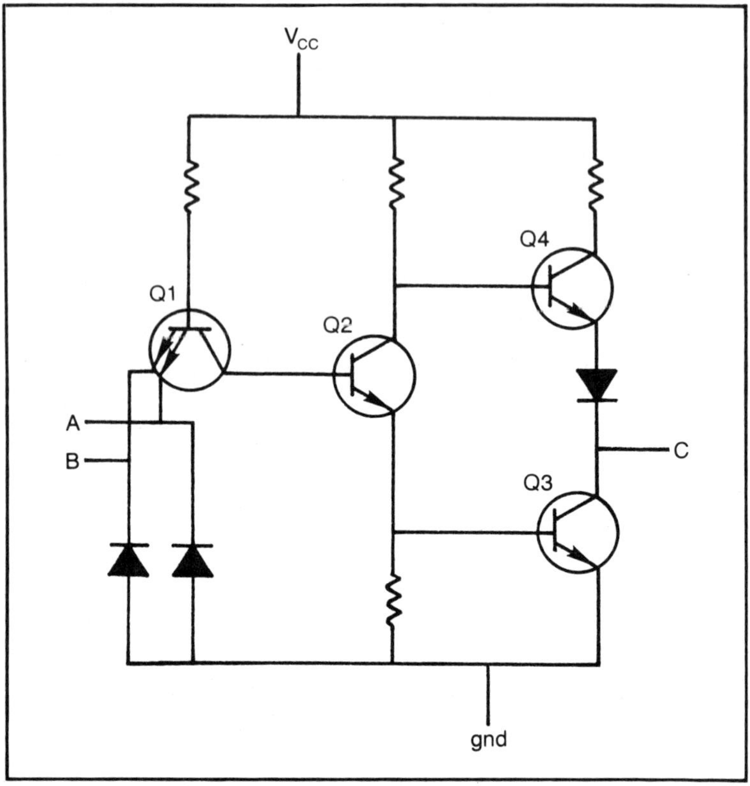

Fig. 4-14. Two-input TTL NAND gate.

up in the common base configuration, and biased so that if one of the emitters is low (grounded), the transistor will be on. The diodes on the input are to prevent a negative voltage from damaging the device.

If one of the emitters is grounded, the collector of Q2 will be almost at ground, turning it off. This is because of the reverse bias on Q2 because of the emitter resistor. If Q2 is off, Q3 will be off because it's base will be at ground through the Q2 emitter resistor. Also, if Q2 is off, the collector will be at Vcc, biasing Q4 on. If Q4 is on and Q3 is off, the output will be high. The result is that the output will be a 1 or high whenever one or more of the inputs is a 0 or low.

When both inputs are high, Q1 is not conducting. Therefore, the base of Q2 is biased high through the base collector junction of Q1. If Q2 is on, there is collector to emitter current, and voltage drops across both the emitter and collector resistors. The voltage drop across the emitter resistor will forward bias the base of Q3,

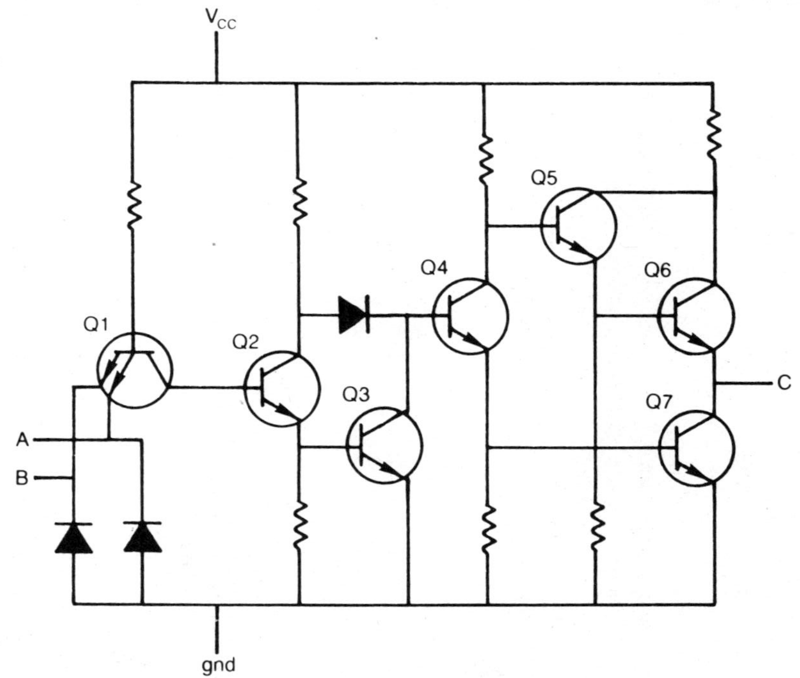

Fig. 4-15. Two-input TTL NAND gate.

turning it on. The voltage drop across the collector resistor is sufficient to reverse bias the base of Q4, turning it off. The result is that Q3 is on and Q4 is off, giving a low output for all inputs high. Thus, the conditions for a NAND gate are met.

The two input AND gate is shown in Fig. 4-15. It is basically the same as the NAND gate, with a few transistors added to provide inversion. It uses the same type of multiple input transistor for the input. When both inputs are high, there is no emitter current in Q1. This results in a high output by virtue of Q7 being off and Q6 being on. Any low input turns Q1 on, and consequently Q7 is turned on and Q6 turned off. Thus, the requirements for an AND gate are met. That is, when both inputs are high, the output is high, and when one of the inputs is low, the output is low.

These circuits illustrate the use of transistors in TTL logic circuits. Transistors are easy to fabricate as part of an IC. In fact, they are easier to fabricate than resistors because they consist of pn

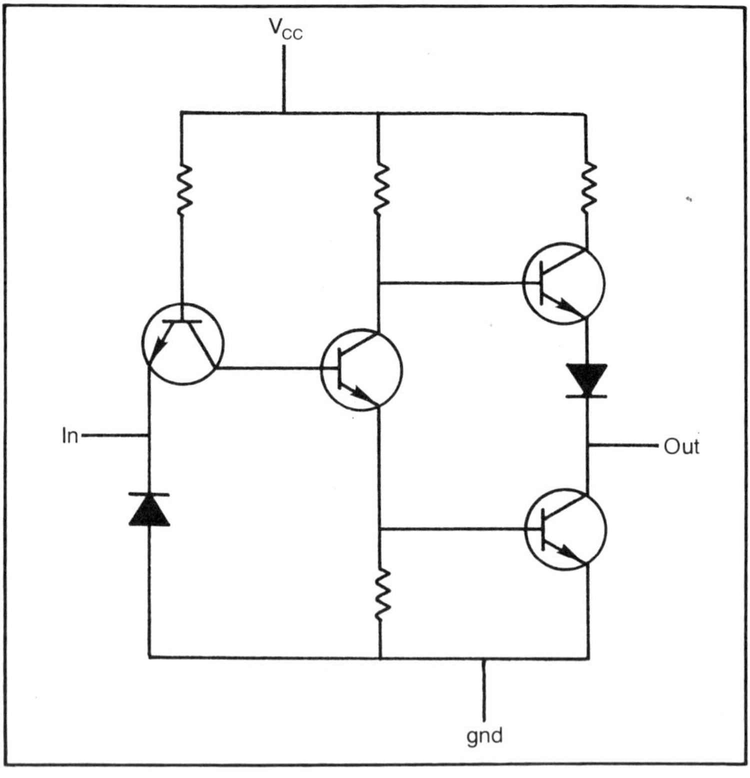

Fig. 4-16. TTL inverter circuit.

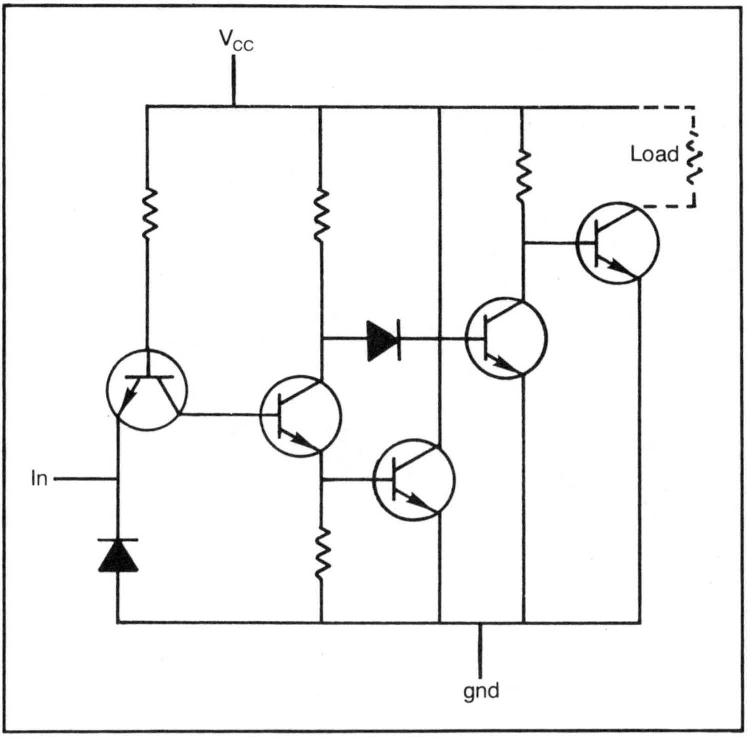

Fig. 4-17. Open collector inverter circuit.

junctions in lieu of conductive material laid out in a pattern. They cost no more in material cost, and they require less design time. The complexity of the circuit is not a factor in the cost once the initial investment has been recovered.

The inverter is a NAND circuit with one input. This is as shown in Fig. 4-16. Notice that it is the same circuit as the NAND gate in Fig. 4-14, except that it has only one input. The buffer, or driver is typically an AND gate with only one input.

There is a class of driver circuits available. These are open-collector circuits which can drive higher voltage and/or higher current loads. Figure 4-17 shows a typical open-collector inverter. The output voltage and current are limited by the output transistor, and for at least one type of open-collector inverter, this is 30 mA output low current instead of the typical 16mA for TTL 7400 series ICs. The load must be referenced to a voltage which can be up to 30 volts, instead of the typical 5 volts. But the load must be connected between the output and the voltage, as shown by the dashes.

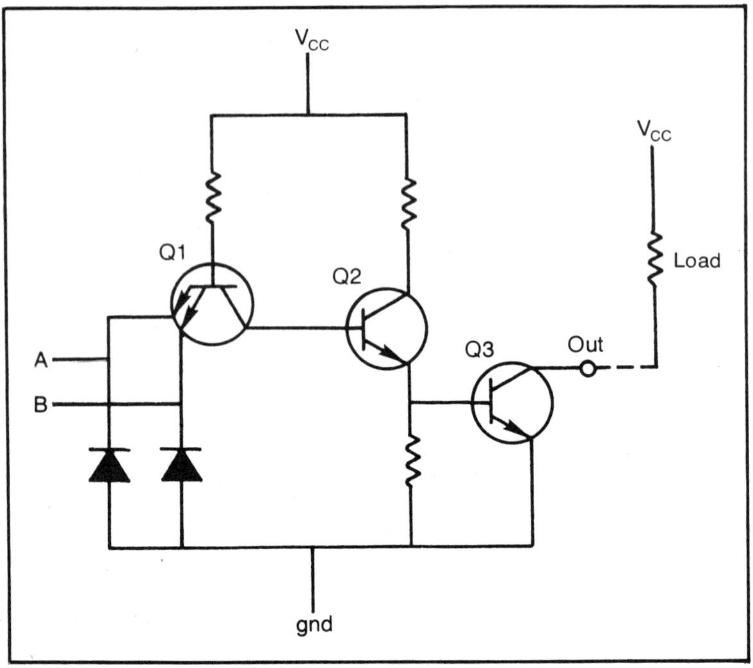

Fig. 4-18. Open collector two-input NAND gate.

Open-collector outputs are available in many gates, such as the AND, NAND, inverter, and buffer. Figure 4-18 shows an open-collector NAND gate. Notice the similarity of this figure with the NAND gate shown in Fig. 4-14. They are the same, except that Q4 and it's associated diode and resistor have been omitted.

There are two approaches to making a NOR gate. Referring to Fig. 4-19, if input A is high, Q1 is not carrying emitter current. Therefore, Q2 is conducting current because it is biased on by the voltage through the collector-to-base junction of Q1. If Q2 is conducting current, Q5 is on and Q6 is off, giving a low output. The result is that the output is low for a high input.

If both inputs are low, both Q1 and Q2 are on. Then Q2 and Q3 are biased off by the ground applied through the collector to emitter of the input transistors. This results in Q6 being on, Q5 being off, and a high output.

If one of the parallel transistors, Q2 and Q3 are on, there will be collector-to-emitter current, making for a low output. One transistor being on determines the results. Therefore, more inputs can be added by adding input transistors and parallel transistors.

The second approach, as shown in Fig. 4-20, uses parallel control transistors to control the state of the output transistors, as in the above circuit. Notice that Q4, Q5, and Q6 are in parallel and control transistors Q7 and Q8. If one of the parallel transistors is on, Q8 will be on and Q7 will be off, resulting in a low output. If all the parallel transistors are off, Q8 will be off and Q7 will be on, resulting in a high output.

If all the inputs are low, transistors Q1, Q2, and Q3 are conducting current. This results in Q4, Q5, and Q6 being off because the base is clamped to ground through the collector to emitter of the input transistors. As shown above, if all the parallel transistors are off, the output is high. This results in a high output if all the inputs are low.

If one input (input A) is high, the corresponding input transistor (Q1) will be off. This is because there is no current path for the

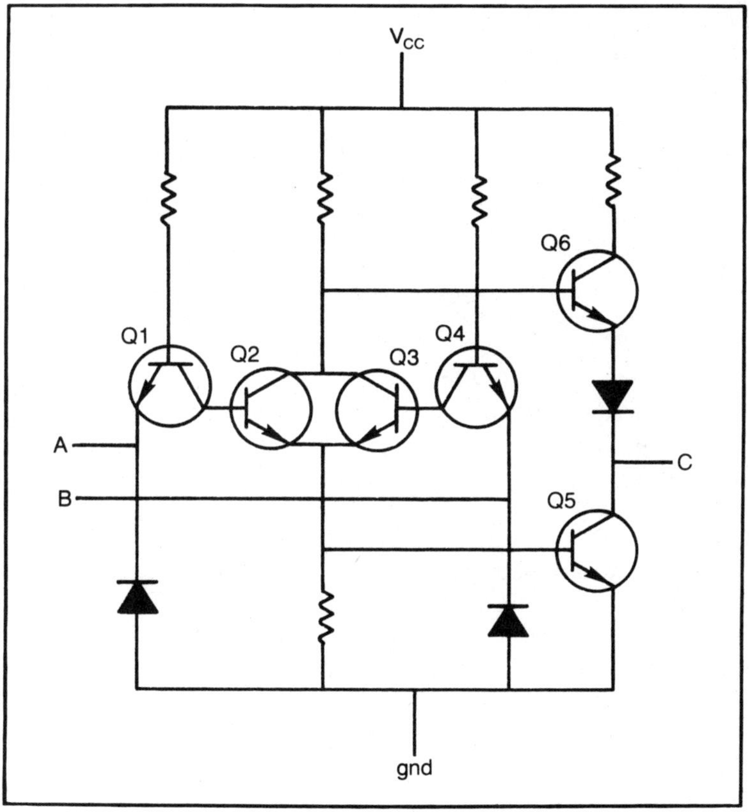

Fig. 4-19. Two-input TTL NOR gate.

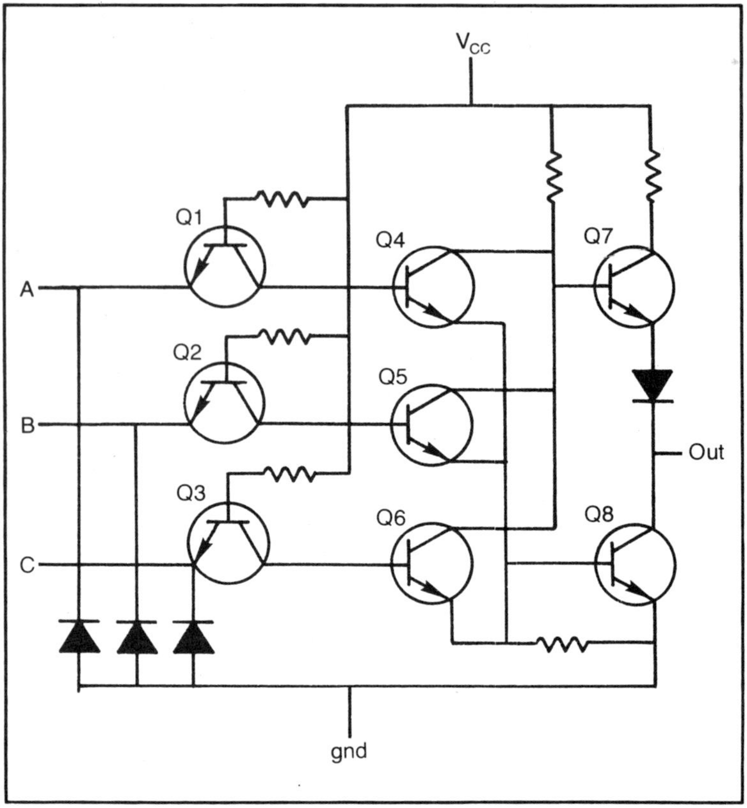

Fig. 4-20. Three-input TTL NOR gate showing an alternate circuit configuration.

emitter current. If Q1 is off, the base of Q4 is biased through the base-to-collector junction of Q1, turning it on. This transistor being on results in a low output. The result is that if any of the inputs are high, the output will be low. Thus, this circuit meets the requirements for a NOR gate.

Most OR gates are similar with the addition of transistor stages to invert the output. Figure 4-21 shows one such gate. It is a two-input version of Fig. 4-20 with the addition of Q5 and Q6 to provide inversion. If desired, the reader can go through the circuit in the manner used above to determine the output for the various inputs.

Using IC technology, some TTL ICs are combinations of the basic functions, such as an AND-NOR gate, but these are more complex. Some of the more advanced ICs are discussed in Chapter 6.

CMOS. CMOS technology, or an offshoot of it, will be the logic circuit of the future. It is finding more and more uses as some of the problems are being overcome. CMOS ICs use *field effect transistors* in place of regular transistors in the switching circuits. FETs are high impedance devices that operate with extremely low currents. Although these characteristics are an advantage, they can also be a problem. This makes them sensitive to some types of electrical noise and to electrostatic discharges.

CMOS ICs can operate over a wide variation of voltage inputs and power. For instance, a typical CMOS digital gate can operate from 3 V to 15 V, while a typical 7400 series TTL IC operates from 4.5 V to 5.5 V. In addition, the quiescent current is several orders of magnitude less. This makes the power supply requirements for a CMOS circuit considerably less. The switching levels are a function of the power supply voltage, and the higher the voltage the wider apart the 0 level and the 1 level are. This allows for a wider margin of safety.

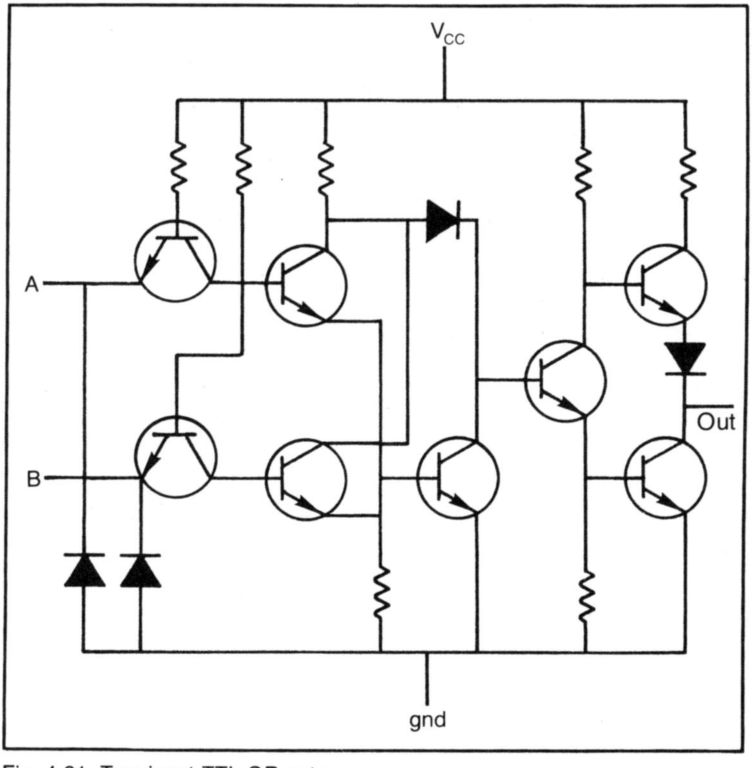

Fig. 4-21. Two-input TTL OR gate.

The FET is a voltage driven device and has high impedance. Therefore, they can operate directly from the power supply without current limiting. This reduces the parts count for the inactive components in the IC, while increasing the active component count. But it is easier to make several identical FETs than it is to make some FETs and some resistors. So the price of the CMOS digital logic circuits is decreasing and will soon be in the area of the TTL circuits.

Figure 4-22 shows some of the schematic diagrams for the FET. There are many configurations for the FET, and only a few of them are given. Figure 4-23 shows the schematic for a 2-input NAND gate. Notice that there are no passive components, only FETs. The utilization of the active components is maximized.

There are CMOS equivalents for most of the 7400 series gates. These are pin and function equivalents, but they cannot be used interchangeably with the 7400 series unless the differences are accounted for. The CMOS series is identified as the 74C series, with the C standing for CMOS. There are also digital CMOS ICs which are not correlated to any standard logic circuit.

ANALOG

Analog circuits operate between the full on and full off conditions. They usually generate a varying output voltage which is a function of the input with a rate of change from zero (dc voltage) to a high frequency. Any output level generated by the circuit is stable, and a

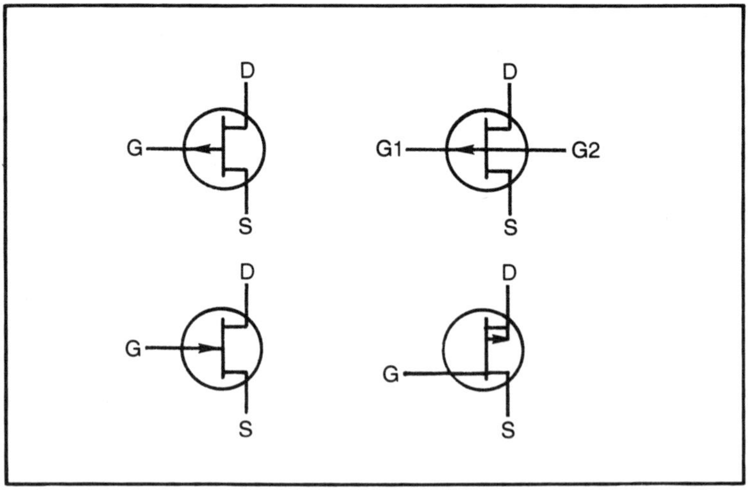

Fig. 4-22. Some circuit symbols for the FET.

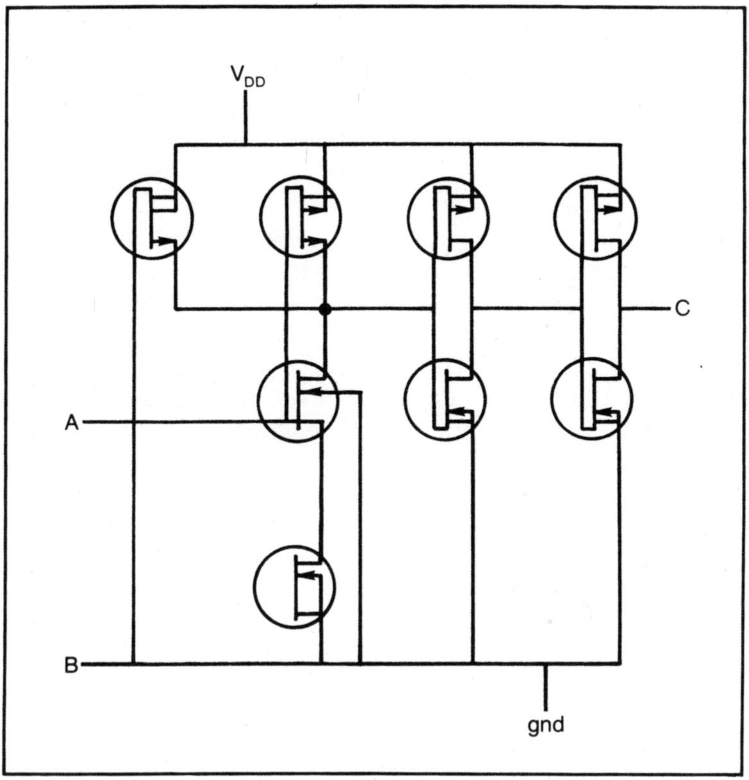

Fig. 4-23. CMOS two-input NAND gate.

function of the input. These conditions apply most of the time, but there are analog circuits which have only a high and a low level for outputs. The main differentiator between analog and digital circuits is that digital circuits operate using switching circuits of some type.

Analog circuits are also called *linear circuits* because they operate the transistors in the linear region of the response curve. This is the region where the output is a linear function of the input.

There are literally hundreds of different functions available as analog ICs. Some of the simpler functions are listed here:

Analog switches
Voltage regulators
Operational amplifiers
Audio Amplifiers
Timers
Analog-to-digital converters

Digital-to-analog converters
RF amplifiers

The list goes on and on. Of the functions available, we are concerned with the first five of the functions. These represent a cross section of those available, and will be studied in detail to illustrate the operation and use of analog ICs.

Most analog circuits are treated as black boxes with the output given as a function of the input. This eliminates the need to understand the circuitry inside the IC, and relies on the ability to understand the function. Some of the analog ICs become quite complex, and understanding their circuit operation is difficult. For the simpler analog ICs, the circuits are discussed in this chapter. For the more complex circuits, the circuits are only discussed in general.

Analog Switches

The simplest of the analog ICs to be covered is the analog switch. This CMOS device is an electronic on/off switch. Either the switch is open or closed. The part number for the type we will talk about is the MC14066B. The block diagram for this integrated circuit is shown in Fig. 4-25. There are four independent switches with an input, output, and control. This is a bidirectional switch that can carry current from the input to the output, or from the output to the input. This switch can connect either a current source or a current sink to the load.

Figure 4-26 shows the circuit diagram for one of the four switches in the package. Since it is a CMOS device it uses field effect transistors for the active elements. The diodes on the input,

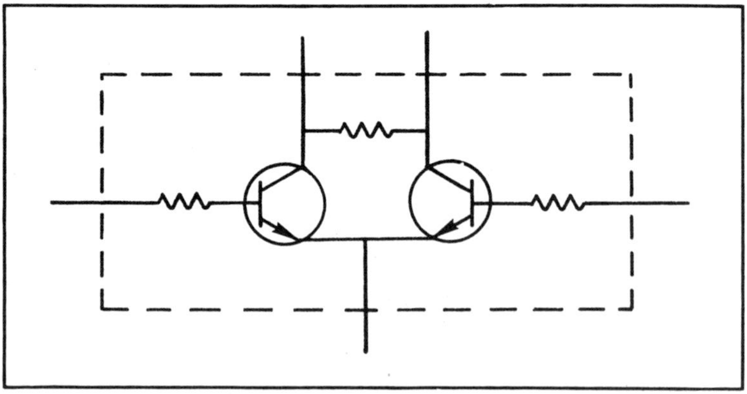

Fig. 4-24. Early analog IC composed of a differential amplifier.

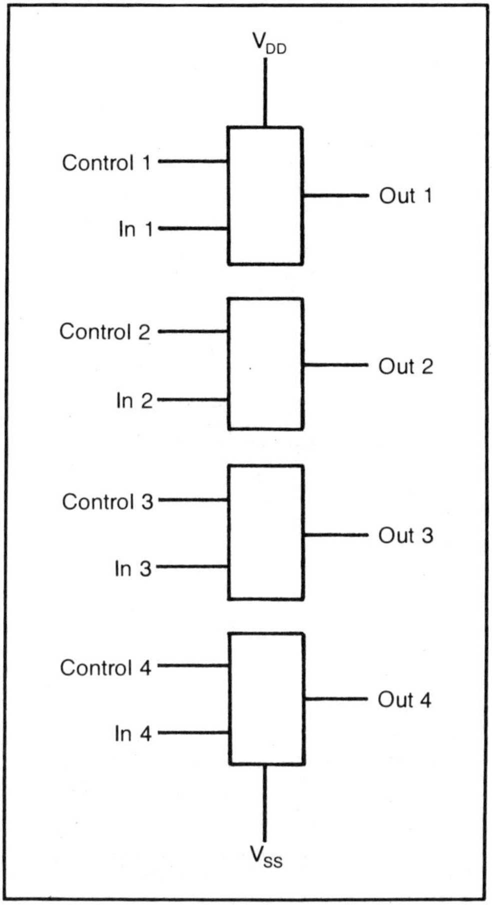

Fig. 4-25. Analog switch block diagram.

output, and control are for circuit protection. The resistors limit the current in case high voltage is applied.

Without going through FET theory, it is sufficient to say that when the control voltage is high, the switch is closed, and when the control voltage is low the switch is open. When the switch is closed, there is an input-to-output resistance because of the circuit elements. This resistance is typically 300 ohms or less depending on the type number. When the switch is open, the input-to-output resistance is extremely high.

When the circuit is enabled, or on, the voltage at the input pin appears at the output pin. This voltage may be anywhere between VDD (the applied power supply voltage), and VSS (typically ground). Since this circuit is a bidirectional switch, any voltage at the output

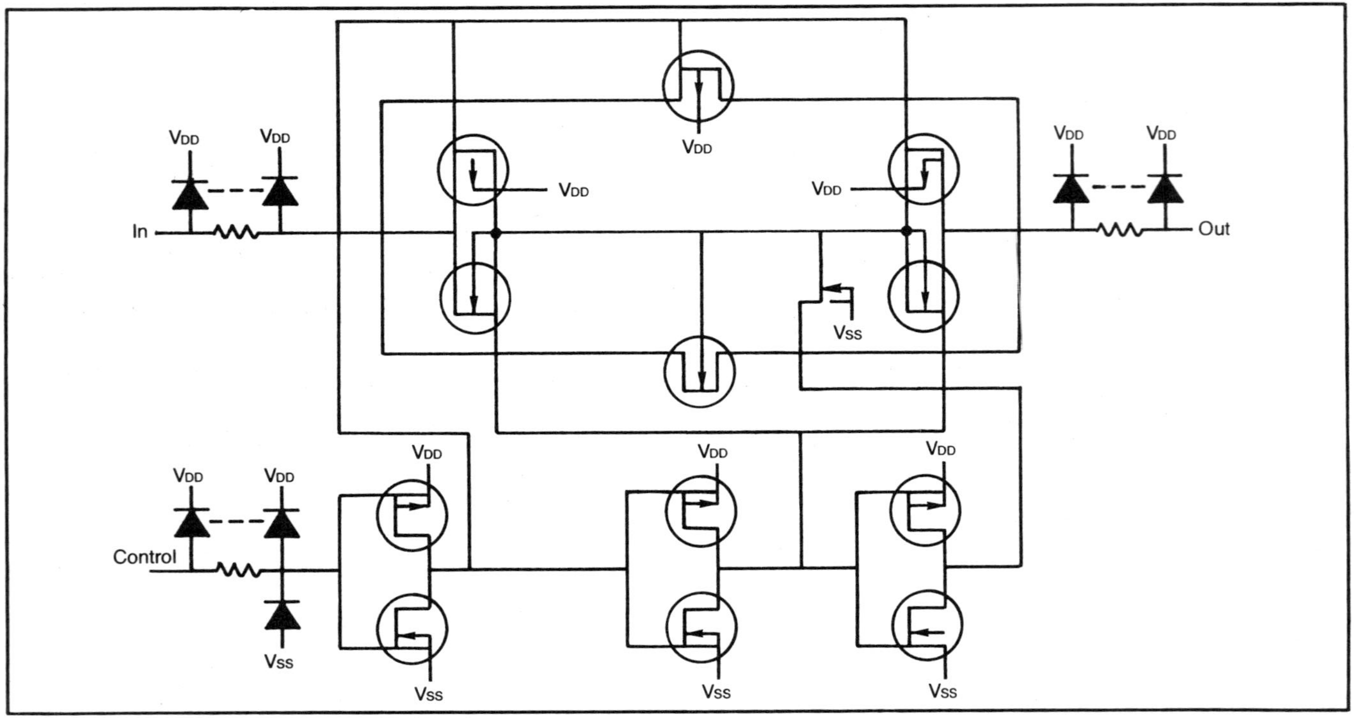

Fig. 4-26. Circuit diagram for one of the analog switches.

will also appear at the input. So if voltage is applied to both pins by the external circuitry, the resultant voltage may not be what is desired. Care must be taken to assure that this condition does not happen when using bidirectional switches.

Analog switches are used for many switching applications. These include telemetry and data collection where several channels of data are multiplexed into one channel. Another widespread use is in automatic testing. The analog switches are used to connect the test signals to the unit under test. Many control circuits use analog switches to route the control signals to the proper end point.

The analog switch is usually represented by a rectangle with the input, output, and control lines labeled. Switches come in many configurations, each with it's own applications.

Voltage Regulators

Voltage regulators consist of a class of integrated circuits which convert unregulated dc voltage to a lower regulated dc voltage level. The voltage regulator integrated circuit has greatly simplified the design and building of low current dc power supplies. The regulator circuit does not have to be designed and built with discrete components. Instead, the single IC is used, which is considerably cheaper and takes less space. A transistorized voltage regulator that I built takes three transistors, two zener diodes, several resistors, and a couple capacitors.

The function of a voltage regulator is to isolate the load from the dc power source so that changes in current draw do not affect the voltage. In simplified form, the regulator is a series transistor with control circuitry. The control circuitry controls the bias on the base of the transistor to change the voltage drop across the transistor so that the output voltage is constant. This is illustrated by Fig. 4-27. For such a circuit, V_{in} is typically the output of the diode rectifiers of the power supply.

There are two types of regulation which are used to determine the quality of a power supply or regulator. These are load and line regulation. *Line regulation* is the change in the output voltage for a change in the input voltage.

$$R_{line} = \Delta V_{out} / \Delta V_{in}$$

It may be expressed as percent, or as millivolts change per volt input. In essence, line regulation is the ability of the regulator to overcome changes in input voltages and hold the output constant.

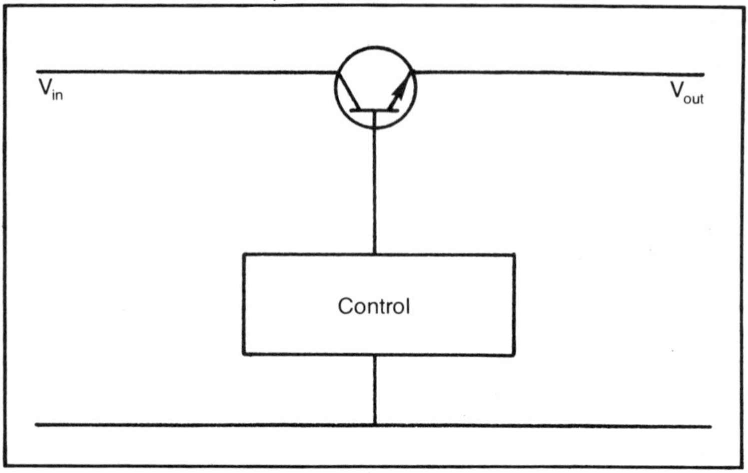

Fig. 4-27. Simplified regulator diagram showing series transistor and control circuit.

Load regulation is the ability of the regulator to maintain a constant output voltage for changing load conditions. It is the change in output voltage for a change in output current.

$$R_{load} = \Delta V_{out} / \Delta I_{out}$$

It is usually expressed as millivolts change per amp change.

There are three types of integrated circuit regulators available.

- ☐ Three terminal fixed voltage—The output voltage is fixed by the integrated circuit.
- ☐ Three terminal adjustable voltage—The output voltage is determined by a voltage divider external circuit.
- ☐ Multipin adjustable voltage—Several external components are required to establish the voltage and current. This was the first type to appear. It has been replaced by the above types.

Of these types, the first is the most popular. The output voltage is determined by the IC and is reflected in the part number. For example, the LM340T5 regulator discussed in this book is the 5 volt version of the LM340T family of regulators. This family contains 8 different regulators with a different output voltage and a different last number in the type number.

Figure 4-28 shows the fixed voltage three terminal regulator on the left and the adjustable three terminal regulator on the right. The three terminal regulator has three terminals: input, output, and ground. As their name implies, they are used for the input voltage, the output voltage, and ground respectively.

The internal circuitry consists of 15 to 20 transistors, numerous resistors, some diodes, and maybe even some capacitors depending on the voltage regulator type number. The internal circuitry is quite complex, and will not be discussed here. Typical IC regulators provide additional functions not provided by the standard discrete component regulator.

☐ Overcurrent protection—This circuitry turns off the regulator if the output is short circuited to ground. The regulator turns back on when the short is removed.
☐ Thermal protection—This function reduces internal power dissipation or turns off the regulator when the IC gets hot. This prevents thermal damage to the regulator.
☐ Overvoltage protection—This function turns off the output in event of an internal failure.

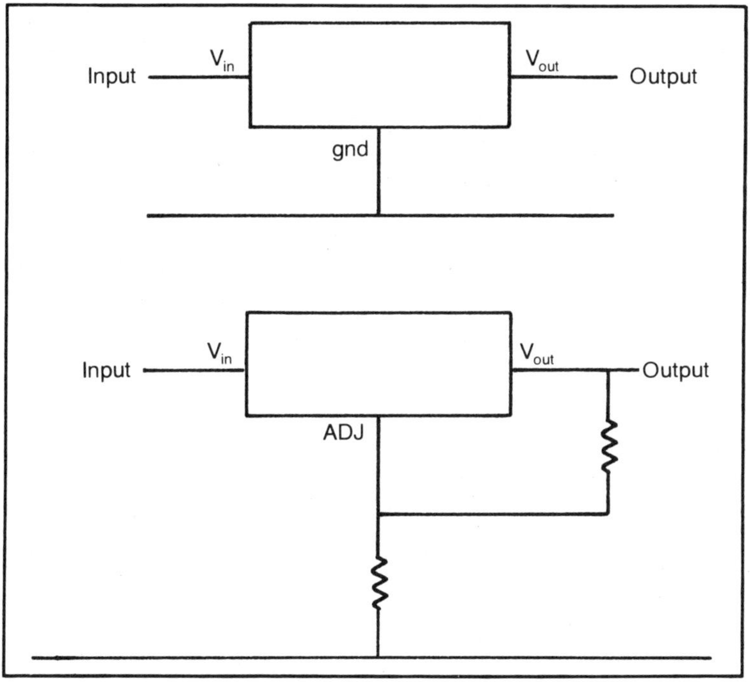

Fig. 4-28. Fixed and adjustable three-terminal regulators.

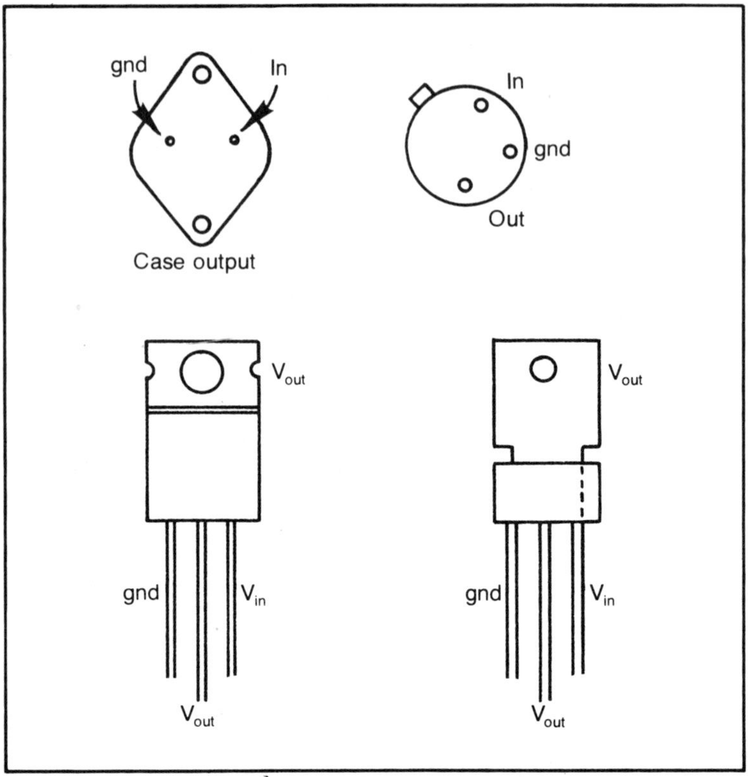

Fig. 4-29. Typical case diagrams for three-terminal regulators.

Integrated circuit voltage regulators usually look like power transistors, with three connections and a heatsink attachment. Figure 4-29 gives the typical outline drawings and typical pin assignments. The pin assignments may be different for the various type numbers.

Operational Amplifiers

Operational amplifiers, sometimes called op amps for short, are high gain amplifier circuits with a differential input. External components are used to define the function of the circuit. Figure 4-30 shows the schematic symbol for an op amp. The + and – symbols designate the positive input and the negative input. The output is the line at the point of the triangle, opposite the inputs. Some operational amplifiers require external stability and compensation circuit. If this is the case, these leads are shown in the upper or lower sloping sides.

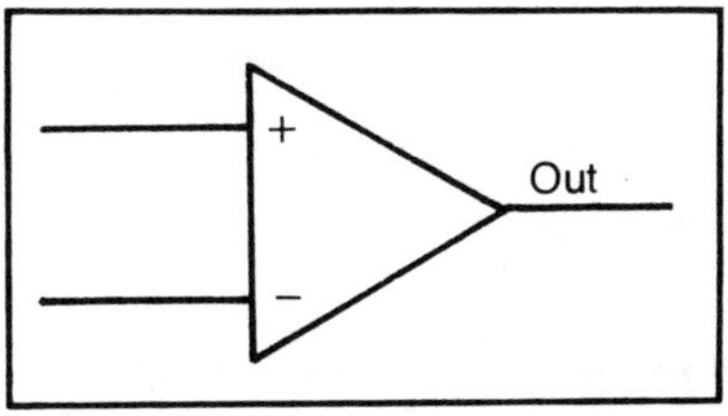

Fig. 4-30. Operational amplifier functional symbol.

The operational amplifier uses a differential input amplifier as the input stage. This is illustrated in Fig. 4-31. This is an emitter coupled amplifier whose output is from output 1 to output 2, and is the difference between the two input signals. The output is not directly referenced to ground unless one of the inputs is referenced to ground. The differential amplifier stage used matched transistors to eliminate the common mode error signals. There may be more than one differential amplifier stage, but eventually the signal is referenced to the common of the amplifier and a single line output is generated.

Figure 4-32 shows the combination of two signals by a differential amplifier. Since the output is the difference between the two inputs, it is input 1 minus input 2. Input 2 is inverted then added to input 1, giving the indicated results.

Since there is no capacitive coupling in the operational amplifier circuitry, the frequency response is a function of the frequency response of the elements used in the IC. This ranges

Fig. 4-31. Basic differential amplifier.

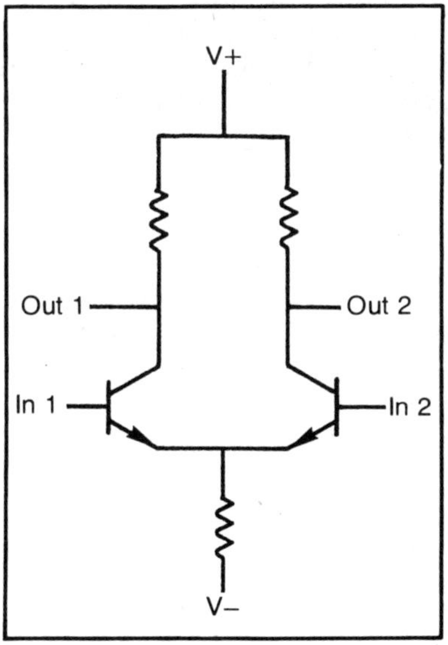

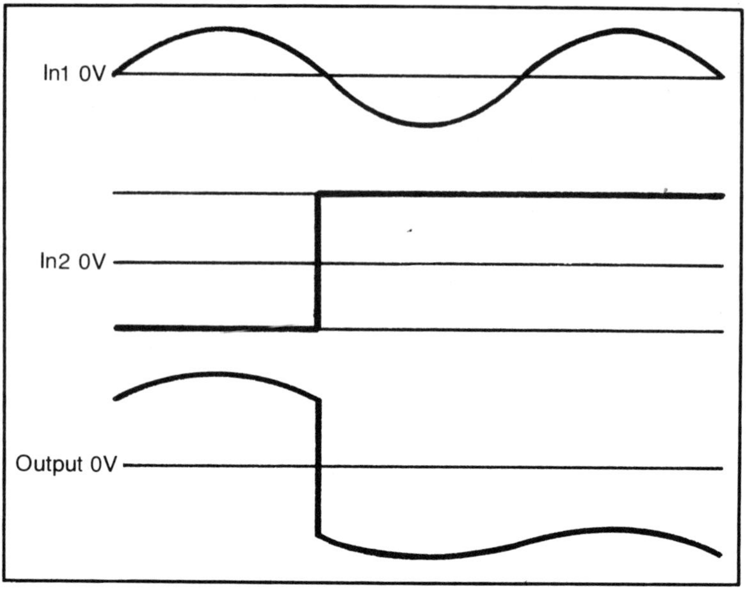

Fig. 4-32. Waveshapes for differential amplification.

from dc to several megacycles. This is further expanded in the next chapter on specifications.

The op amp has a + and – input. A signal applied to the + input will not be inverted, while a signal applied to the – input will be inverted. This means simply that the polarity of the output is the same as the + input, and opposite that of the – input. To establish the function of an op amp circuit, external components are connected between the + input, – input, and the output. Since the op amp has high gain, it is unstable without some type of feedback. It will do all sorts of strange things, like saturate the output and oscillate. Feedback is used to stabilize the circuit.

Figure 4-33 shows a simple operational amplifier circuit. It is connected as an inverting amplifier, because the input is to the minus input. The gain of this circuit is defined by the following equation:

$$\text{gain} = -R_f/R_i$$

Where R_f is the feedback resistance and R_i is the series input resistance. The minus sign indicates that the output signal is inverted. This is a very common op amp circuit that is used as an amplifier circuit. Chapter 7 covers more applications.

The LM324 operational amplifier, as discussed in the following chapters, is a general purpose op amp. It has good open circuit gain. Good op amp circuit design dictates that some of the standard methods be used to insure circuit stability. These are covered in a later chapter.

Operational amplifiers can either be powered from a single power supply, or from two power supplies. For most op amps, the single supply results in the output being above zero volts. Using a positive and a negative supply results in the operating point being at zero volts, and the output fluctuating both positive and negative.

Audio Amplifiers

Audio amplifiers, sometimes called power amplifiers, are integrated circuits designed for use as audio amplifiers and preamplifiers. The fidelity and frequency response is not good enough for these to be considered hi-fi circuits. As the circuit design improves, and better processes are developed, the response will be that of a hi-fi circuit.

Preamplifiers usually follow low-level output transducers such as microphones, phonograph pickup devices, and tape read heads. These are all low output devices, so the amplifier must have a low noise figure and match the impedance of the transducer. The noise figure is a measure of the amplifier to reject noise at the input and keep internally generated electrical noise at a minimum so that it is not detectable at the output. Electrical noise is unwanted electrical signals generated by the circuit components or picked up by the

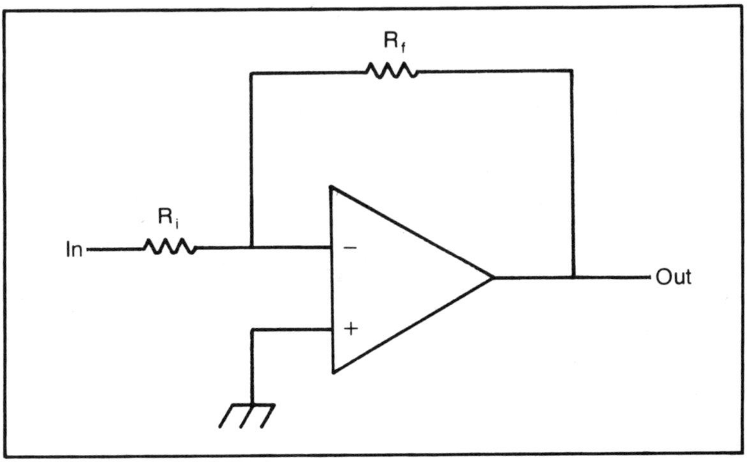

Fig. 4-33. Simple operational amplifier circuit.

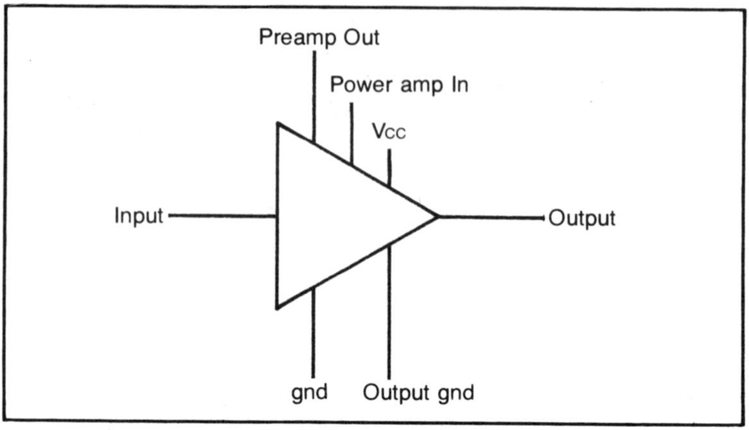

Fig. 4-34. Functional symbol for power amplifier IC.

circuit. It usually shows up as static or as bursts of noise. Any noise generated or amplified by the preamplifier may be amplified enough to distort the output. This is especially critical for amplifiers with a very low input signal level. These amplifiers must take noise into account. Since the signal input for a power amplifier is several orders of magnitude greater than the internally generated noise, the noise figure is not as important for power amplifiers.

Figure 4-34 gives the functional diagram for the MC1306 power amplifier as discussed in this book. This power amplifier

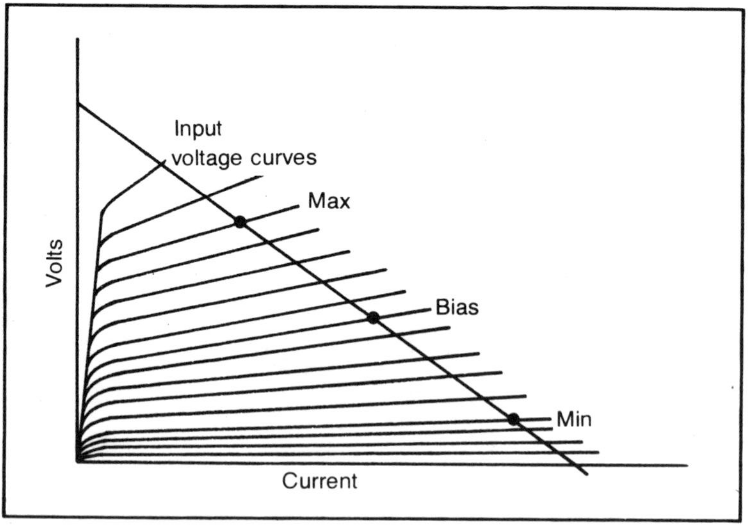

Fig. 4-35. Class A amplifier response curve.

contains both a preamplifier and a power amplifier. The input on the left side is the preamplifier input. Notice that there is a preamplifier output and a power amplifier input along the top. The output on the right side is the power amplifier output, and this is connected to the coupling circuit for the speaker. There is a separate ground for the power amplifier to allow direct connection to the output transistor, and it is normally connected to circuit ground.

The MC1306 power amplifier integrated circuit is a ½ watt audio amplifier containing about 14 transistors, less than a dozen resistors, and some diodes. It will drive an 8 ohm speaker from a 12 volt power source. Other power amplifier ICs are available which will drive a 2 ohm or a 4 ohm speaker load with up to 12 watts. Most of these have internal preamplifiers.

Timers

The timer integrated circuit was chosen as a representative of a group of analog ICs classified as industrial functional blocks. This timer IC is usually represented by a functional block, as shown in Fig. 4-36, with the inputs and outputs labeled as to function and pin number. Sometimes the function is not given, and must be looked up. The pin numbers may be in numerical order, but most of the time they are arranged to facilitate drawing the external components in the circuit diagram.

A timer circuit operates in one condition for a predetermined

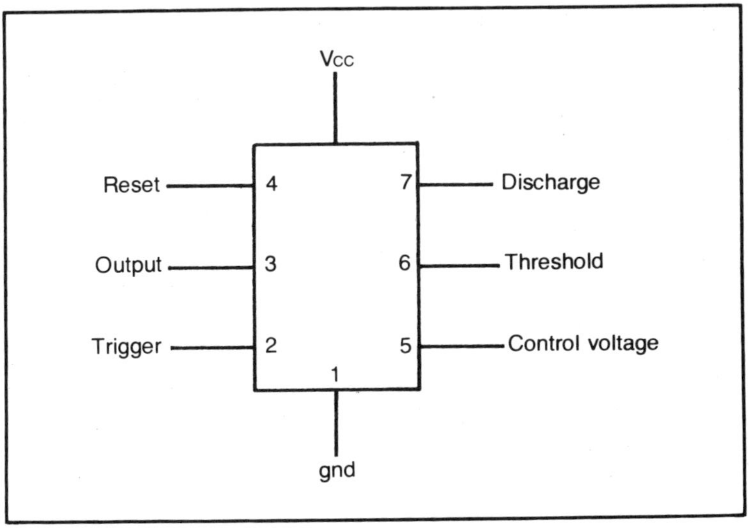

Fig. 4-36. Functional symbol for the 555 timer IC.

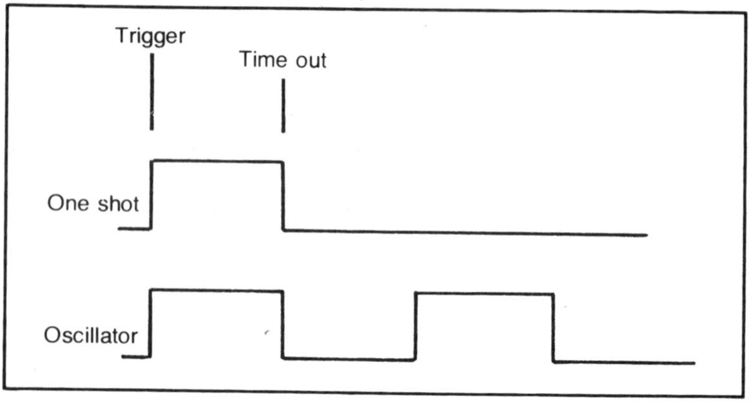

Fig. 4-37. One shot and oscillator waveshapes.

length of time, then switches states. Typically, when they are triggered, the output changes to the unstable condition. When the timing circuit times out, the output switches back to the stable condition, generating one pulse. This is illustrated by the top waveshape of Fig. 4-37, which is the response for a one-shot circuit. A one shot generates one pulse of a predetermined length of time. The time is a function of the external resistor and capacitor timing components. For the 555 timer IC discussed in this book, this time can be from microseconds to hours, depending on the value of the timing resistor and capacitor.

Most timing circuits can be connected as free-running oscillators. That is, they produce a series of pulses without external triggering. The lower waveshape of Fig. 4-37 shows the output of an oscillator. The ratio of the time the output is high divided by the time for one pulse, both high and low output, is the *duty cycle* of the output. It is expressed in percent. The high time, low time, and duty cycle are all controlled by the resistor and capacitor timing network. The details of determining the values required are given in a later chapter.

The 555 timer is relatively easy to use and only requires a few resistors and two capacitors. While the output is stable enough for most operations, there is a tendency for the output to be pulled slightly off frequency under some load conditions. Isolation of the timer from the load by a buffer will help this condition.

5

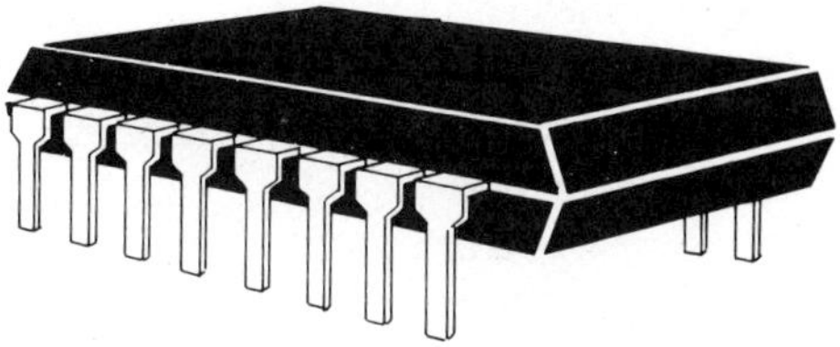

Specifications

Now that we have looked at some of the ICs as an introduction to the circuits, it is time to look at the specifications and what they mean. In this chapter we will look for the specifications for the integrated circuits that will be discussed in detail in chapters 6 and 7, and used in the projects in chapter 8.

The simple 7400 series gates are covered first, then the analog circuits, then the complex 7400 series TTL integrated circuits.

The specifications for an integrated circuit cover many things needed to use the IC.

Power supply voltage range
Power supply current
Input voltage range
Input current range
Output voltage
Output current
Response waveshapes
Absolute maximum ratings
Pin diagram

These specifications are originated by the manufacturer of the IC and may be determined by analysis or by actual measurements. Some of the specifications are tested for during the manufacture of the device, while others are assumed. Some of the specifications give a minimum number, others give a maximum number, while others give a typical number. Some give typical values and either a maximum or minimum. The typical number is what can be expected from a majority of the devices. The maximum or minimum values allow determining the worst case condition.

Integrated circuits come in packages, and all the ICs covered in this chapter except the voltage regulator are packaged in dual in-line packages (DIP). The standard DIP package has pins on 0.1 inch centers along the opposite long sides. The parallel rows of pins are 0.3 inches apart. Figure 5-1 shows the top view for an 8-pin, 16-pin and 14-pin DIP packages. There is an index mark of some type at one end, and the pin numbering starts from this index. For the top view, the numbering is counterclockwise, with half the pins on each side. When the pins are turned up, the pin numbering is clockwise. When connecting up a circuit, the pin numbering must be correct.

Of all the specifications available for integrated circuits, only those required for operation within the scope of this book are given and discussed. When a specification is first encountered it is discussed in detail, after that it is just mentioned.

The 7400 series digital IC specifications are similar and are discussed first under one grouping. Then the analog ICs are discussed one at a time.

Many IC manufacturers have available individual specification sheets for the ICs. Most of them also have specification books. These books each cover one grouping of ICs, such as TTL, Linear, Voltage Regulators, CMOS, Audio Amplifiers, etc.

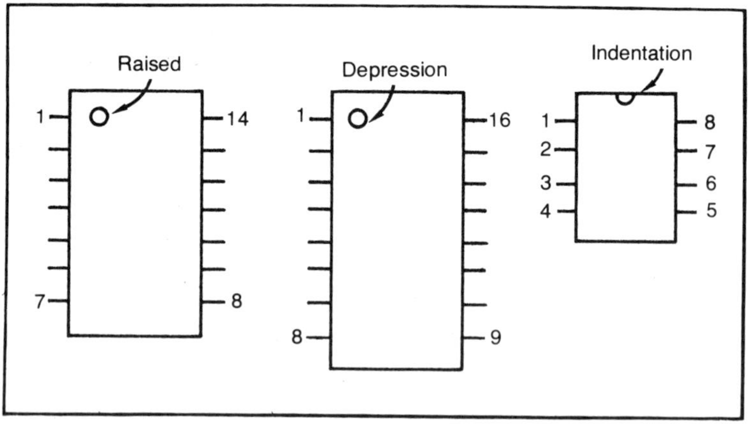

Fig. 5-1. The top view of IC DIP packages showing pin numbering and index mark.

SIMPLE 7400 DEVICES

All of the 7400 series TTL digital ICs discussed in this chapter have the same absolute maximum ratings. These ratings are given in Table 5-1. The absolute maximum ratings are those values which must not be exceeded. Exceeding these maximum ratings may permanently damage the integrated circuit. The parameters are as follows:

☐ Supply Voltage—The absolute maximum value which the applied supply voltage must not exceed. The IC may not operate at this voltage.
☐ Input Voltage—The absolute maximum voltage at the input pins. This applies to any one input pin, no matter what the condition of the other input pins are. This is the absolute maximum logic level for a logical 1 or high.
☐ Operating Temperature—This is the temperature range over which the circuit will operate.
☐ Storage Temperature—This is the range of temperatures which the IC will survive in the nonoperating condition. Exposing the IC to temperatures outside this range may physically damage the device.

The inputs for TTL devices are protected against negative input voltages by input clamping diodes. These are good for about 12 mA, and serve as protection against negative transients. The logical 0 or low state must never be allowed to go negative, or below ground.

Figure 5-2 shows the pin numbering and logic diagrams for the 7400, 7402, and 7408 integrated circuits. Note that each of these ICs is a quad gate, which means that each package contains four independent gates. For simple logic functions, the logic diagram is usually given with the pin functional diagram. For complex functions, the function of each pin is usually given. The pin diagrams are given as top views, looking down on the IC with the pins down, as it

Table 5-1. Absolute Maximum Ratings for the TTL 7400 Series Logic ICs

Supply Voltage, V_{CC}	7 V
Input Voltage, V_{in}	5.5 V
Operating Temperature	0 °C to 70 °C
Storage Temperature	−65 °C to 150 °C

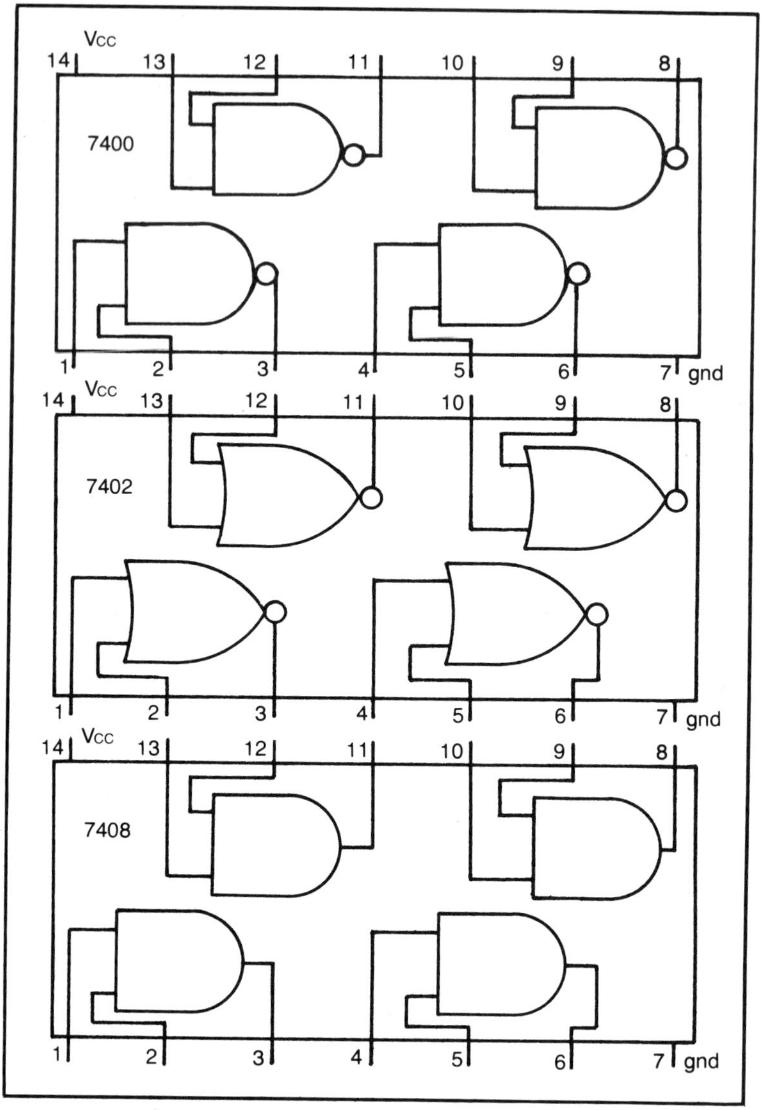

Fig. 5-2. Pin functions and logic diagrams for the 7400, 7402, and 7408 digital gates.

would look in a circuit. All of these ICs are 14-pin DIP packages with pin 7 as ground and pin 14 as VCC, or the power supply connection.

Some of the more complex 7400 series ICs have 16-pins. Usually pin 16 is VCC and pin 8 is ground, but there are exceptions. Always check the pin diagram.

Table 5-2 gives the dc characteristics for the three digital ICs in Fig. 5-2. Note that they are almost all the same, and these values are typical for the 7400 series simple gates. The explanation of each parameter is given here:

- V_{IL} — Low level input voltage. This is the maximum voltage level for a logic 0 for this IC to recognize it as a 0.
- V_{IH} — High level input voltage. A logic 1 must be at least this value to be recognized as a logic 1.
- V_{OL} — Low level output voltage. A logic 0 output will be the typical value for most of the devices, and no device will exceed the maximum value.
- V_{OH} — High level output voltage. A logic 1 will be at least the typical value for most of the devices. No devices will be below the minimum value. The output loading will affect this value.
- I_{IL} — Low level input current. This is the maximum input current for a 0 input.
- I_{IH} — High level input current. This is the maximum input current required to assure a logic 1 input.
- I_{OL} — Low level output current. This is a current sink capability for a logic 0 output. This is also the drive capability of the output.
- I_{OH} — High level output current. This is the minimum current available at the output for a 1 state.

Table 5-2. DC Characteristics for the 7400, 7402, and 7408 Digital ICs

Parameter	7400	7402	7408	Units
V_{IL} max	0.8	0.8	0.8	Vdc
V_{IH} min	2	2	2	Vdc
V_{OL} typ	0.22	0.22	0.22	Vdc
max	0.4	0.4	0.4	Vdc
V_{OH} min	2.4	2.4	2.4	Vdc
typ	3.3	3.3	3.3	Vdc
I_{IL} max	−1.6	−1.6	−1.6	mAdc
I_{IH} max	40	40	40	μAdc
I_{OL} min	16	16	16	mAdc
I_{OH} min	−400	−400	−400	μAdc
I_{OS} min	−18	−18	−18	mAdc
max	−55	−55	−55	mAdc
I_{CCL} typ	12	14	18	mAdc
max	22	27	26	mAdc
I_{CCH} typ	4	8	10	mAdc
max	8	16	15	mAdc

I_{OS}— Output current, short circuit. This is the minimum and maximum current at the output with the output shorted to ground.

I_{CCL}— Power supply current, low level. This is the power supply current drawn by each gate when the output is at a logic 0 or low level.

I_{CCH}— Power supply current, high level. This is the power supply current drawn by each gate when the output is at a logic 1, or high level.

The negative values for some of the currents indicates that current is drawn from the IC to the external circuit. This can be either at the input or output. Going back to the circuit diagrams for typical TTL gates, a low input grounds the emitter of the input transistor, resulting in current flow from the collector of the input transistor, through the emitter and to the drive circuit. Thus current flows from the IC to the drive circuit.

Notice from Table 5-2 that the current required for a 0 input is many times the current required for a 1 input. Also note that the output current for a 0 is many times that for a 1. This is one of the characteristics of the TTL family. The 0 state carries the most current and is the driven state. For most, but not all, an open input is recognized as a logic 1 input.

The next characteristic to be covered is the *propagation delay* time. This is the times it takes the IC to respond to a changing input. Table 5-3 lists the delay times for the three 7400 digital ICs. These are typical for most of the 7400 series, and limit the frequency of operation of the ICs. More complex digital ICs have more series gates, and therefore may have longer delay times. Figure 5-3 defines the propagation delay times from the applied and response waveshapes.

t_{PLH}—The delay time from the input to the output for an input going from a 0 to a 1 condition.

t_{PHL}—The delay time from the input to the output for an input going from a 1 to a 0 condition.

Table 5-3. Propagation Delay Times for the 7400 Series ICs

Parameter	7400		7402		7408		Unit
	Typ	Max	Typ	Max	Typ	Max	
t_{PLH} (low to high)	11	22	12	22	17.5	27	ns
t_{PHL} (high to low)	7	15	8	15	12	19	ns

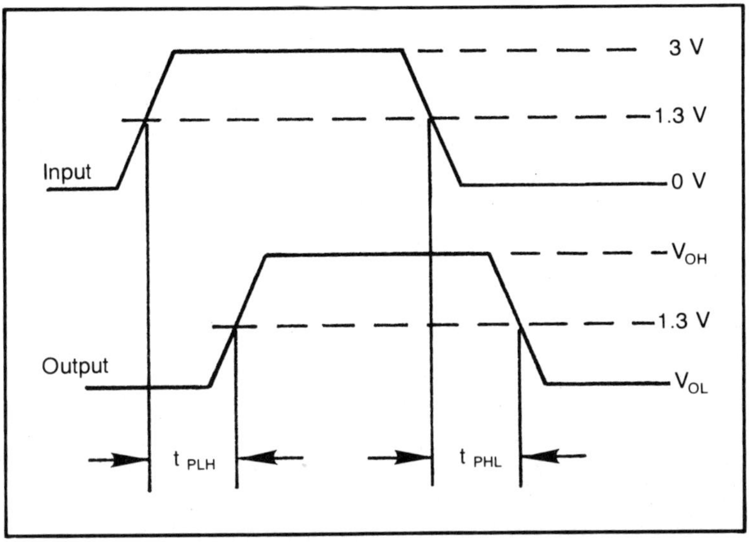

Fig. 5-3. Waveshapes used to define the propagation delay times.

Note for these delay times, the measurements are taken at the 1.3 volt level. This is typically the 50 percent point for the signals during the testing of the ICs. Trying to measure from the 0 point or the 100 percent point is extremely difficult because these points may not be clearly defined on the waveshapes.

The t in these two specifications defines these parameters as ac characteristics. That is, they are conditions of changing states for the device. The test circuit for the propagation delay times usually has a capacitive load of 15 pf and a resistive load of 400 ohms. This simulates a typical circuit application. For drivers and buffers this value may be different.

For the values given in Table 5-2, the gain of the IC can be determined. If the input currents are divided into the output currents, current gain is achieved, as per the following.

$$\text{Gain (1 state)} = I_{OH}/I_{IH} = 400/40 = 10$$
$$\text{Gain (0 state)} = I_{OL}/I_{IL} = 16/1.6 = 10$$

The current gain for the IC is 10. This means that each output can drive up to 10 inputs for both the high and low conditions. If we let the input currents (40 μA and 1.6 mA) define one unit load, then the output can drive 10 unit loads. Typically, this is how the drive capabilities are given, as unit loads. Most 7400 series ICs require one unit load for drive and will drive 10 unit loads. Drivers and buffers can drive 20 unit loads or more. Table 5-4 lists some of the

Table 5-4. Input Drive and Output Current Sink Capability for 7400 Series ICs

IC Type	Function	Unit Loads	
		Drive	Output
7400	NAND	1	10
7402	NOR	1	10
7405	O.C. Inv	1	10
7406	O.C. Inv	1	20
7437	NAND	1	30
7440	NAND	1	30
7460	Expand	1	4

7400 series ICs, their function, the input drive requirements, and the output drive capability.

Sometimes the recommended operating conditions are given. This lists the recommended range of values for the power supply voltage (Vcc) and the 0 and 1 states. For the ICs listed here, and most of the 7400 series, the recommended Vcc is 5 volts with an allowable range of 4.5 volts to 5.5 volts. This does not mean that the power supply can vary over this range during operation. It does mean that the power supply voltage should fall in this range, the closer to 5 volts the better. The range for the 0 state is from ground to V_{IL}, or from 0 volts to 0.8 volts. The 1 state can vary from V_{IH} to Vcc, or from 2 volts to 5 volts.

For optimum switching times and minimum noise response all unused inputs of AND and NAND gates must be tied high. One method of doing this is to connect the unused inputs to used inputs. This can be done if the loading of the drive circuits is not exceeded. Another commonly used method is to connect the unused inputs through a 1 kΩ resistor to Vcc. Several unused inputs can be connected to the same resistor. Both methods are illustrated in Fig. 5-4.

All unused inputs for OR and NOR gates must be connected to ground or in parallel with other inputs. Connecting them to ground does not load the signal and disables that input so that it does not effect the output.

ANALOG DEVICES

Analog integrated circuit specifications give the information required to use the IC in actual circuits. Since these ICs cover many functions, different information is required for the different functions. The specifications may include:

Absolute maximum ratings
Response characteristics
Response curves and waveshapes
dc characteristics
ac characteristics
Pin diagram
Functional diagram
Application information

The information allows the user to determine the inputs required and the resulting outputs. This information is needed for circuit design and also in circuit analysis. The specifications will be examined and discussed, one IC type at a time. When new specifications are encountered they will be defined. The application information is covered in Chapter 7.

The specifications for the same type number, made by different manufacturers, are very similar. Those that affect the circuit operation are usually identical. If there is a doubt, check the manufacturers specifications.

MC14066B Analog Switch

The MC14066B analog switch contains four independant electronic switching circuits capable of controlling either digital or analog signals. Some of the features of this switch are as follows:

Low quiescent current, typically 0.5 nA

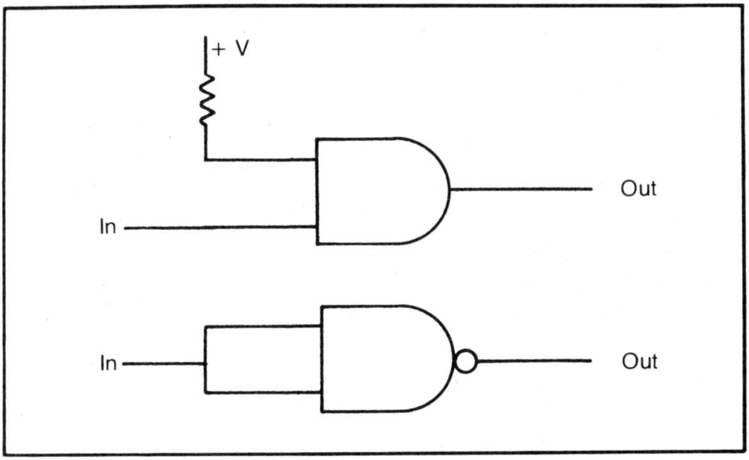

Fig. 5-4. Connections for unused AND and NAND gate inputs.

Low crosstalk between switches
Diode input protection
Wide supply voltage range, from 3 Vdc to 18 Vdc
Wide frequency range, dc to 65 MHz at 10 Vdc
Low internal noise
Easy to use

The quiescent current is the current drawn by the integrated circuit when all the switches are in the open condition. Crosstalk between switches is the signal leakage from one switch to another. This must be extremely low for any integrated circuit with more than one circuit in the package. Internal noise is the electrical noise generated by the IC.

Figure 5-5 shows the pin connections for this IC. The pin definition and function are shown for each of the four switches along with the power and ground connections. Notice from the schematic given in the preceding chapter that the switches are bidirectional and that the input can serve as the output and the output can serve as the input. You must be aware of this fact when using the integrated circuit so that you can take steps to prevent a signal at the output from being passed through to the input. If the switch is off, there is no problem, but if there is a slight timing problem with the control signal, glitches may appear at the input when the switch is turned on.

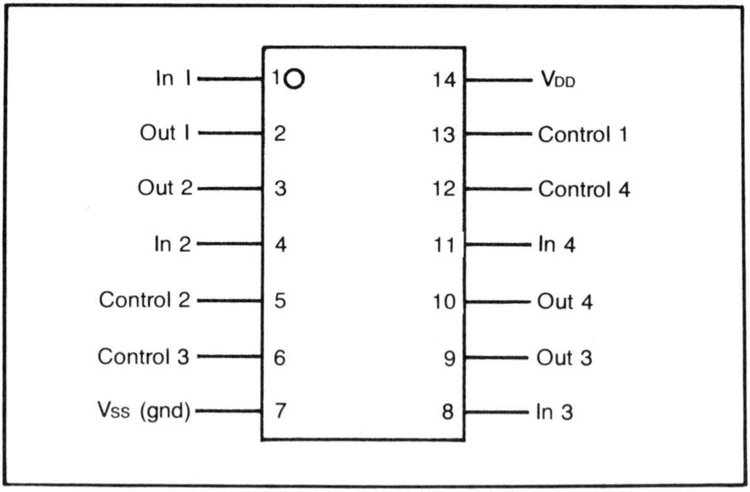

Fig. 5-5. Chip outline diagram for the MC14066B analog switch showing the function of the pins.

Table 5-5. Absolute Maximum Ratings for the MC14066B Analog Switch

Rating	Symbol	Value	Units
dc Supply Voltage	V_{DD}	0 to +18	Vdc
Input Voltage, all inputs	V_{in}	−0.5 to V_{DD} +0.5	Vdc
Through Current	I	25	mAdc
Operating temperature	T_A	−40 to +85	°C
Storage temperature	T_{stg}	−65 to +150	°C

Table 5-5 gives the absolute maximum ratings for the analog switch. VDD is the power supply voltage. It is called VDD because it is the drain voltage for the FETs. VCC is the collector voltage for transistor circuits. Note that VDD can be up to 18 Vdc without damage, but it is advisable to keep it below this value to compensate for power supply drift. Typically, this switch is not operated at higher than 15 Vdc.

The through current is the current the switch is capable of conducting from input to output. Current in excess of this can cause overheating and damage. The typical failure mode for overcurrent is burning holes in the junction by melting small pathways through the semiconductor junction. Each time this happens, the junction is degraded, so the IC should be discarded if current beyond the absolute maximum value is drawn. This applies to semiconductors in general including transistors and integrated circuits.

The input voltage is the voltage at the input pins, and the maximum is referenced to VDD. This is the voltage that can be passed from the input to the output of the switch.

The electrical characteristics for this device are given in Table 5-6. The first two input voltages are for the control input. This is the input which turns the switch on and off. These are given for different values of VDD, and the typical values are slightly less than one half the VDD value. This says that, at 5 volts VDD, the typical device will recognize 2.25 volts as a 0 input, and that all devices will recognize 1.5 volts as a 0 input. So it is advisable to keep the 0 level control input below the limit value for VDD. At 5 volts VDD, the control input must be at least 3.5 volts for all units to recognize the input as a 1. Note that the input current for the control input is typically in the nanoamp region. So both conditions, on and off, must be driven. Leaving the control pin open can cause random switching.

The *on-resistance* is the resistance between the signal input and output when the switch is on. Note that the typical values can be between 80 and 250 ohms, and are dependent on the value of VDD. If

Table 5-6. Electrical Characteristics for the MC14066B Analog Switch

Characteristic	V_{DD}	Typ	Limit	Unit
Input Voltage (control) 0 level	5.0 10 15	2.25 4.5 6.75	1.5 3 3.75	Vdc
Input Voltage (control) 1 level	5.0 10 15	2.75 5.5 8.25	3.5 7.0 11.25	
Input Current (control)	15	.00001	0.1	μAdc
Input Capacitance (control) (switch)	10 10	5 8	7.5 15	pF pF
Output Capacitance	10	8		pF
On Resistance	5.0 10 15	250 120 80	1050 500 280	Ohms

Table 5-7. Switching Characteristics for the MC14066B Analog Switch

Characteristic	Symbol	V_{DD}	Typ	Max	Unit
Propagation Delay Times Input to Output	t_{PLH}, t_{PHL}	5.0 10 15	20 10 7	40 20 15	ns
Control to Output Output 1 to High Imp.	t_{PHZ}	5.0 10 15	40 35 30	80 70 60	ns
Output 0 to High Imp.	t_{PLZ}	5.0 10 15	40 35 30	80 70 60	ns
High Imp. to Output 1	t_{PZH}	5.0 10 15	60 20 15	120 40 30	ns
High Imp. to Output 0	t_{PZL}	5.0 10 15	60 20 15	120 40 30	ns
Frequency Response (on)		5.0	65		MHz
Max Control Frequency		5.0 10 15	6 8 8.5		MHz MHz MHz

the voltage drop through the switch is critical, this on resistance must be taken into account.

Table 5-7 gives the switching characteristics. These are the times it takes for the switch to change states under various conditions. The first propagation delay time is the time it takes a pulse input signal to appear at the output for both positive and negative going pulses. This is illustrated in Fig. 5-6, which shows the waveshapes which define the various propagation delay times. The remaining four delay times are the time it takes to turn on and off the switch by the control input for both a 1 input and a 0 input.

The frequency response is the maximum frequency that can be passed from input to the output for an on switch. The maximum control frequency is the maximum frequency which the control can be switched at and have the device switch on and off. This is considerably lower than the frequency response.

Table 5-8 gives the control characteristics for the analog switch. A 0 input on the control will turn the switch off, and the switch will have greater than 1000 megohms resistance between the input and output. A 1 on the control input will turn the switch on.

From the specifications, this analog switch will operate at standard TTL logic levels and power supply voltage level. The input logic levels for the control input are compatible with the TTL logic levels.

555 Timer

The 555 timer is a stable integrated circuit useful for generating pulses in the one-shot or oscillator mode. The pulses may be from microseconds to hours long with a few external components. Some of the features are given here:

Timing from microseconds to hours
Operates in both the one-shot and free running modes
Adjustable duty cycle
Output can source or sink up to 200 mA
Input and output TTL compatible

Table 5-8. Control Response for the MC14066B Analog Switch

Control	$V_{control}$	Switch	V_{in} to V_{out} Resistance
0	0 V_{dc}	Off	Greater than 10^9 ohms
1	V_{DD}	On	300 ohms or less

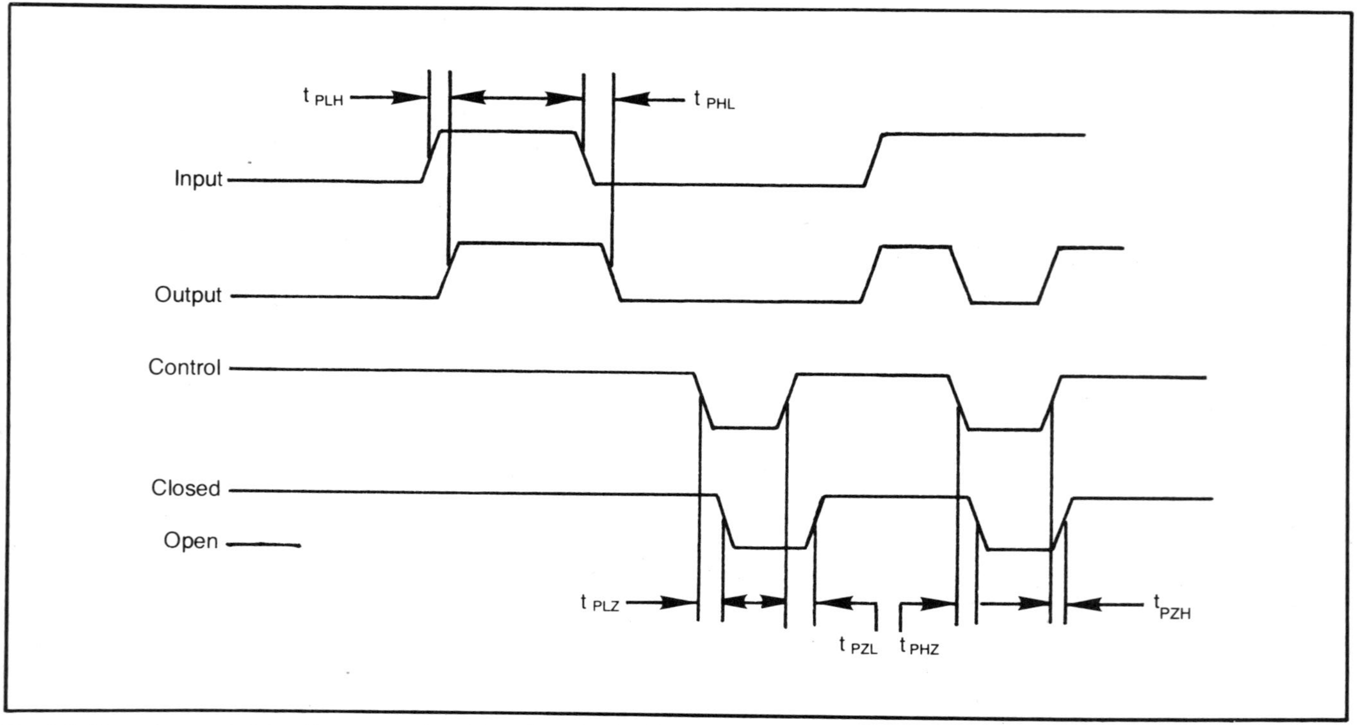

Fig. 5-6. Waveshapes defining the propagation delay times for the MC14066B analog switch.

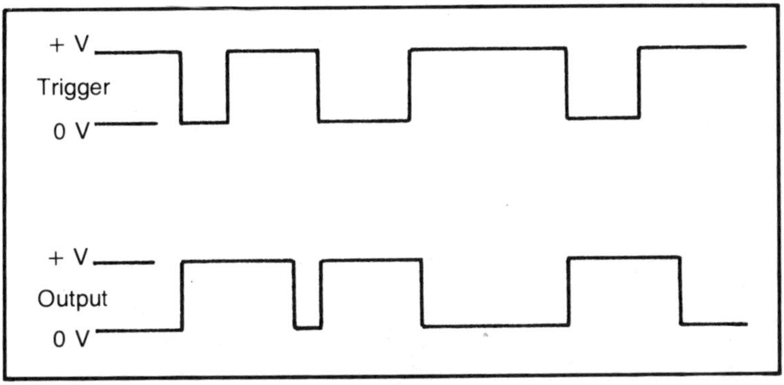

Fig. 5-7. Trigger and output waveshapes for a one shot showing the starting of the output pulse on the falling edge of the trigger.

The basic operation of the integrated circuit is as a one-shot, producing a pulse on the negative going edge of the trigger. The operation, as a one shot is illustrated in Fig. 5-7. This figure shows the output response for the given trigger input. When the trigger goes from + V to 0 V, the one shot triggers, changing the output state from a low to a high. When the timing circuit times out, the output switches back to a low. The timing circuit for the one shot consists of one resistor and one capacitor.

If the trigger signal goes negative again before the one shot has timed out, the timing will start again. So if the trigger repetition rate is faster than the one shot pulse time, the circuit will stay in the triggered state, and will not time out. This capability is used as a missing pulse detector, where one pulse will not let the one shot time out, but two time periods will if there is no trigger in the second time period.

The IC can also be used in the free running, or oscillator mode. In this mode, the output continually changes state, with the time periods being determined by the timing components. This is illustrated in Fig. 5-8, which shows two free running oscillator waveshapes. Note that there are two times involved, they are the up time and the down time. These are defined as the times between the 50 percent points on the positive going and the negative going transitions of the output. The *duty cycle* is defined as the percent of time that the waveshape is high and can be calculated with the following equation:

Duty Cycle= Up time/(Up time+ down time)

Figure 5-8 shows waveshapes for a 50 percent duty cycle and a 33

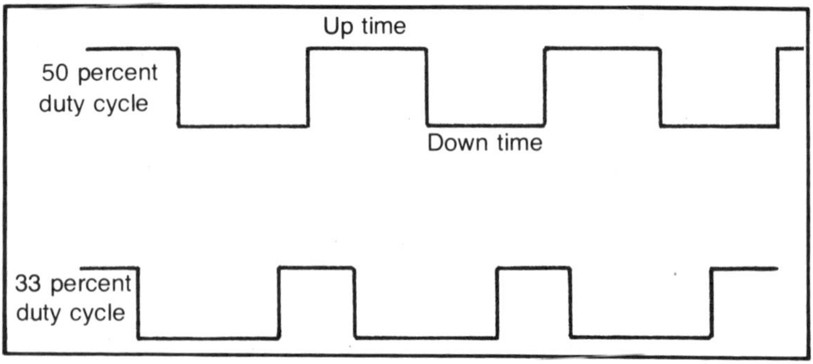

Fig. 5-8. Oscillator output waveshapes for 50 percent and 33 percent duty cycle.

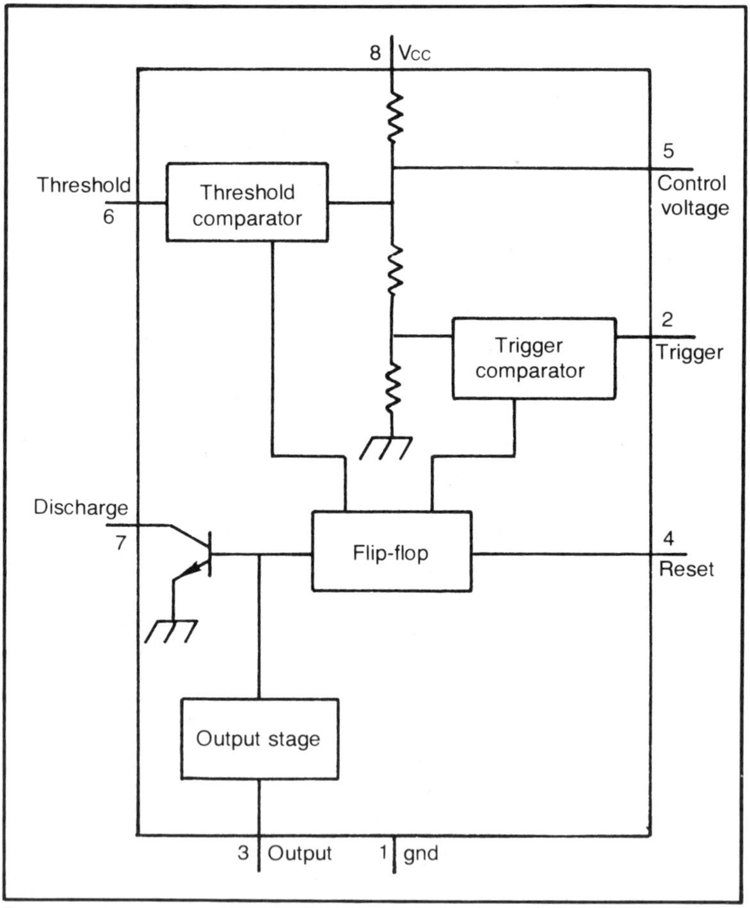

Fig. 5-9. Block diagram and pin functional definition for the 555 timer.

percent duty cycle. Different applications require different duty cycles.

Figure 5-9 shows the block diagram and pin definitions for the 555 timer. It is packaged in an 8-pin DIP package. It is also available in an 8-pin round metal can with the same pin functions.

The reset input is provided to discharge the timing capacitor, thus interrupting the timing cycle. As long as this pin is low, the capacitor is prevented from charging. When this function is not required for circuit operation, it should be connected to V_{CC}.

The discharge input is to discharge the timing capacitor after the one shot has timed out. This conditions the timing components for the next trigger input. Since this pin is connected to the junction of the timing resistor and capacitor, the timing resistor must provide adequate current limiting.

As shown in the block diagram of Fig. 5-9, the internal circuitry consists of two comparators, a flip-flop, a reset circuit, and an output circuit. The threshold comparator determines when to time out the circuit. The threshold input samples the capacitor charge voltage from the timing circuit. The control voltage input provides an external means of setting the trigger level for the threshold comparator. If this control input is not used, it is normally connected to ground through a 0.01 μF capacitor to prevent electrical noise interaction. If the control input is not used, the trigger level is set by the resistor divider network. When the threshold voltage equals the control voltage (either internal or external), the threshold comparator terminates the output pulse.

The trigger comparator compares the trigger input against an internally generated reference voltage. When the trigger input equals this value and is negative going, the trigger comparator initiates the pulse and timing cycle.

The flip-flop controls the output stage and discharge transistor. A flip-flop is a circuit which has two stable states, on and off. It will be in one state until it is switched to the other state by one of the inputs. The flip-flop turns the output on when the trigger comparator switches the flip-flop on. It will remain on until switched off by the threshold comparator. This turns off the output stage and turns the discharge transistor on.

Table 5-9 gives the absolute maximum ratings for this IC. The power dissipation is the power that can be dissipated by the IC without overheating or damage. This is concerned with keeping the chip temperature below 150°C, and the main contributors are the current load and the discharge current. Normally, if the output

Table 5-9. Absolute Maximum Ratings for the 555 Timer

Rating	Value	Units
Supply Voltage	+18 V	volts
Power Dissipation	600 mW	milliwatts
Operating Temp.	0 °C to +70 °C	degrees C
Storage Temp.	−65 °C to +150 °C	degrees C
Lead Temperature	300 °C (10 seconds)	degrees C

current is kept to a few milliamps, this does not become a problem. If in doubt, touch the IC while operating, if it is hot, chances are it is overheating.

The lead temperature is the temperature which the leads can withstand. There is usually a time associated with this temperature. This specification is mainly concerned with the temperature when soldering the IC into a circuit. The heat is conducted through the lead into the package and to the chip itself. The 10 seconds given for this specification does not mean 10 seconds on one lead, followed immediately by 10 seconds on another lead. It means that after 10 seconds the chip must be given a chance to cool down. If sockets are used for the ICs, lead overheating is avoided. Soldering is discussed in more detail in the chapter on projects.

Table 5-10 gives the electrical characteristics for this device. The supply voltage for this IC can be from 4.5 to 18 volts, and it is usually operated with a supply voltage from 5 volts to 15 volts. The capability of operating from a 5 volt power supply means that it can be powered by the power supply when used with TTL logic circuits. Although the voltage at the capacitor of the timing circuit, and subsequently, the threshold voltage, is a function of the supply voltage, so is the internal control voltage. The result is that the pulse time is relatively independent of the supply voltage.

The supply current is a function of the supply voltage for the low state, as is expected. This is because the low state is the driven state while the high state is the idle state.

The threshold voltage and current, reset voltage and current, trigger voltage and current, and the control voltage are all given. These values establish the requirements for these inputs if they are used. The threshold is two-thirds of V_{CC} while the trigger is about one-third of V_{CC}. These will be discussed in more detail in Chapter 7.

340T-5 Voltage Regulator

The 340T-5 is a typical three terminal voltage regulator. It is basically a dc-to-dc converter which adds several degrees of regulation. This device is a member of the 340 series of regulators, which offer several dash numbers with several different output voltages. Some of the features of this device are:

One amp capability
Line regulation of 3 mV/volt, typical
Load regulation of 10 mV/amp, typical
Internal thermal overload protection
Internal short-circuit protection
Output transistor protection

Internal thermal overload protection means that the internal circuitry will decrease the output current and even shut the device off if excessive internal heat is generated. Internal short-circuit protection means that the device will turn off if excessive current is drawn by the output. These two features required additional circuitry for discrete component power supplies.

Table 5-10. Electrical Characteristics for the 555 Timer

Parameter	Min	Typ	Max	Units
Supply Voltage	4.5		18	Volts
Supply Current Low state: Vcc = 5 V Vcc = 15 V High state		 3 10 1	 5 12	 mA mA mA
Threshold Voltage		.667		× Vcc
Trigger Voltage: Vcc = 5 V Vcc = 15 V	1.45 4.8	1.67 5	1.9 5.2	Volts Volts
Trigger current		0.01	0.5	μA
Reset Voltage	0.4	0.5	1	Volts
Reset Current		0.1	0.4	mA
Threshold Current		0.1	0.25	μA
Control Voltage: Vcc = 5 V Vcc = 15 V	2.9 9.6	3.33 10	3.8 10.4	Volts Volts

Table 5-11 gives the absolute maximum ratings for this voltage regulator. The maximum input voltage for the 5 volt regulator is 35 volts. This is considerably higher than required for proper circuit operation, and will severely limit the current available from the power supply. This is because of the internal power dissipation in the device. The main factor in the internal power dissipation is the power dissipated in the output transistor.

$$P_d = (V_{in} - V_{out}) \times I_{load}$$

So if the input voltage is 35 volts and the output is 5 volts, and the load current is ½ amp, the power dissipation becomes:

$$P_d = (35 - 5) \times \tfrac{1}{2} = 30 \times \tfrac{1}{2} = 15 \text{ watts.}$$

Without extremely efficient heatsinking, this device will shut down because of internal thermal overload protection. The thermal calculations are involved and beyond the scope of this book. It is sufficient to say that if the device overheats, use a better heatsink or reduce the input voltage. Some curves for this device connected to some heatsinks are given later.

Table 5-12 gives the electrical characteristics for the voltage regulator. The exact output voltage is determined by the exact values of the internal chip components. The maximum value and minimum value are output values which may be achieved on different chips. Note that the output for one chip may vary between these values.

Line and load regulation were discussed previously, and both are given as millivolts change per volt or amp of change. The typical

Table 5-11. Absolute Maximum Ratings for the LM340T-5 Voltage Regulator

Rating	Value	Units
Input Voltage	35 V	volts
Internal Power Dissipation	internal limit	
Operating Temperature	0 °C to +125 °C	degrees C
Maximum Junction Temp	150 °C	degrees C
Storage Temperature	−65 °C +150 °C	degrees C
Lead Temperature	230 °C	degrees C

Table 5-12. Electrical Characteristics for the LM340T-5 Voltage Regulator

Parameter	Symbol	Min	Typ	Max	Units
Output Voltage	V_o	4.8	5	5.2	volts
Line Regulation Junction Temp = 25 °C V_{in} = 7 to 25 Volts	ΔV_o		3	50	mV
Junction Temp 0 °C to 125 °C V_{in} = 8 to 20 Volts				50	mV
Load Regulation Junction Temp = 25 °C I_o = 5 mA to 1.5 A	ΔV_o		10	50	mV
Junction Temp 0 °C to 125 °C I_o = 5 mA to 1 A				50	mV
Quiescent Current	I_Q		8		mA
Minimum V_{in}	V_{in}	7.3			volts

values assume that the chip temperature is kept below 25°C. The quiescent current is the current drawn by the regulator with no load current drawn.

The minimum input voltage for the regulator to maintain a 5 volt output is given as 7.3 volts. This means that 2.3 volts is required as a series voltage drop across the regulator. This assures that the output transistor can maintain regulation.

Figure 5-10 gives the outline drawing and the pin definitions for the device. The device is a plastic package with a metal tab out the top and three pins out the bottom. The metal tab is connected to the ground pin to facilitate connecting to a heatsink. There is a mounting hole in the metal tab. Although all three terminal regulators have three pins, the functions may not be in the same order. Always check the specification for the particular type number used.

Figure 5-11 gives a graph of the maximum average power dissipation as a function of ambient temperature for three conditions.

No heatsink
10°C/W heatsink
Infinite heatsink

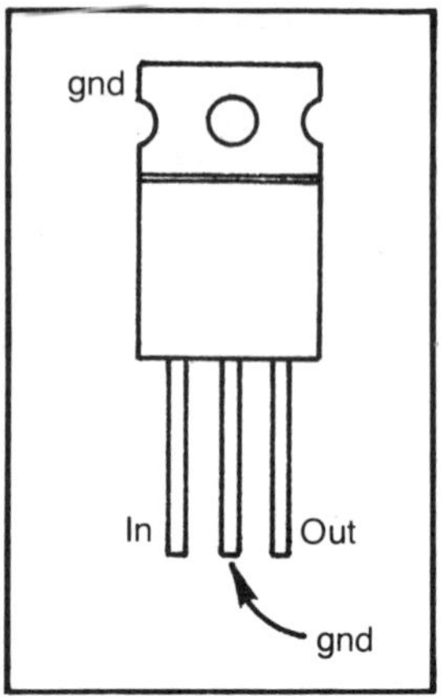

Fig. 5-10. Outline drawing for the voltage regulator.

For no heatsink, the maximum power dissipation is about 1.5 watts, while for an infinite heatsink it is 19 watts. As discussed before, this limits the input voltage and output current.

LM324 Operational Amplifier

The LM324 operational amplifier is a general purpose operational amplifier requiring few external components. Some of the features are:

Four operational amplifiers in a single package
Internal frequency compensation

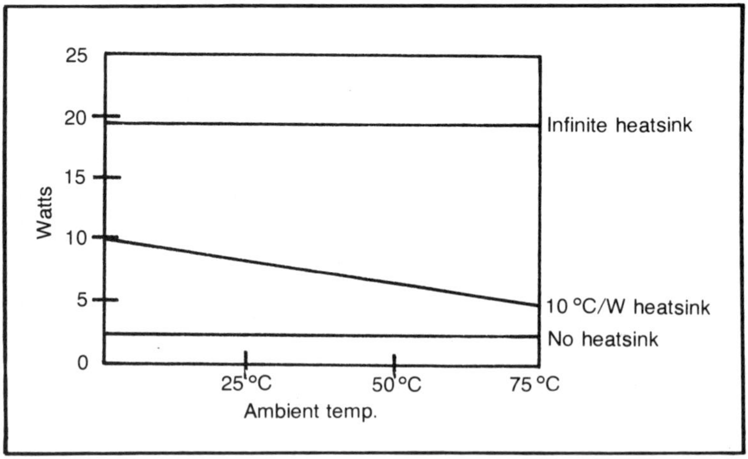

Fig. 5-11. Maximum power dissipation vs ambient temperature for various heat sinks.

Low offset voltage, normally offset adjustment not required
Can operate from a single or dual voltage supply
Wide power supply range
Large dc voltage gain
Large output voltage swing
Compatible with all forms of logic
Low power supply current requirement

This device can be operated from a single 5 volt power supply, or from a supply as high as 30 volts. It requires no offset adjustment for normal operation, and is internally frequency compensated. These features make it easy to use.

Table 5-13 gives the absolute maximum ratings for this operational amplifier. The maximum supply voltage is given for both single and dual supply applications. Single supply operation is when the V+ is connected to a power supply and V– is connected to ground. Dual supply operation is when V+ is connected to the positive terminal of one power supply and V– is connected to the negative terminal of another power supply of the same voltage. The negative side of the positive supply and the positive side of the negative supply are both connected to ground.

The differential input voltage is the voltage at one of the input

Table 5-13. LM324 Operational Amplifier Absolute Maximum Ratings

Rating	Value
Supply Voltage, V+ Single Supply Dual Supply	 32 Vdc ± 16 Vdc
Differential Input Voltage	32 Vdc
Input Voltage	–.3 Vdc to +32 Vdc
Power Dissipation	570 mW
Output Short to Ground	Continuous
Input Current	50 mA
Operating Temperature	0 °C to +70 °C
Storage Temperature	–65 °C to +150 °C
Lead Temperature (10 seconds)	300 °C

pins with respect to the other input pin. This is the normal input for the operational amplifier. The maximum input current will exist only when the voltage at either input pin is driven negative with respect to the voltage at V−. This condition does not normally exist.

Table 5-14 gives the electrical characteristics for the device. Note the large signal voltage gain is 15 volts output for 1 millivolt input, or 15,000. The exact value of this is not significant, as long as it is several times the desired gain of the circuit. The output voltage swing gives the maximum signal output. Note that the upper value approaches V+, and the lower value (for a single supply) approaches ground, within a few millivolts.

Note that the output of this device can source more current than it can sink. This means that it can supply up to 20 milliamps for a high output, but can only sink 8 milliamps for a low output. The input common mode voltage is the differential input voltage, and should be limited to 2 volts less than V+. Otherwise output distortion may occur.

Figure 5-12 gives the functional diagram and pin definitions. Notice that the four individual operational amplifiers are symmetrically connected, with the corner pins being the outputs. The adjacent pins are the minus inputs, and the inside pins are the positive inputs. The minus input is the inverting input, and the positive input is the noninverting input.

Table 5-14. Electrical Characteristics for the LM324 Operational Amplifier

Parameter	Conditions	Min	Typ	Max	Units
Large Signal Voltage Gain	$V^+ = 15$ Vdc R_L Greater than 2 K	15			V/mV
Output Voltage Swing High Out	$V^+ = 30$ Vdc, $R_L = 2$ K R_L greater than 10 K	26 27	 28		Volts Volts
Low out	$V^+ = 5$ V, $R_L = 10$ K		5	20	mV
Output Current Source	$V_{in} = +1$ V, $V^+ = 15$ V	10	20		mA
Sink	$V_{in} = -1$ V, $V^+ = 15$ V	5	8		mA
Short Circuit to Gnd	$T_A = 25$ °C		40	60	mA
Input Common_Mode Voltage	$V^+ = 30$ Vdc	0		$V^+ -2$	Volts
Supply Current	$R_L = \infty$, $V^+ = 30$ V $R_L = \infty$, all amps		1.5 .7	3 1.2	mA mA

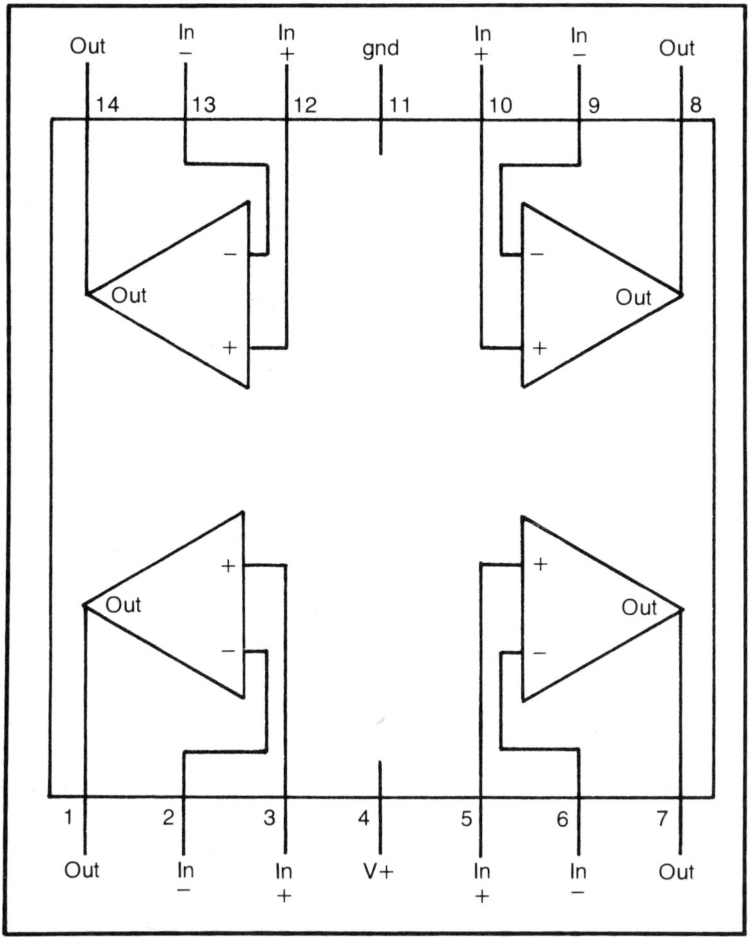

Fig. 5-12. Block diagram and pin functions for the LM324 operational amplifier.

Although pin 11 is labeled as ground, this is the V− connection for dual supply operation. In the dual supply configuration, there is no ground to the device. Ground is established externally and referenced to the inputs.

Figure 5-13 gives the open-loop frequency response along with the circuit used to measure the response. This shows that the higher the gain, the lower the maximum frequency. This curve must be considered when using the operational amplifier for high frequency or high gain applications. Figure 5-14 shows another important frequency response curve. This is the large signal frequency response. The test circuit used to obtain this curve has a

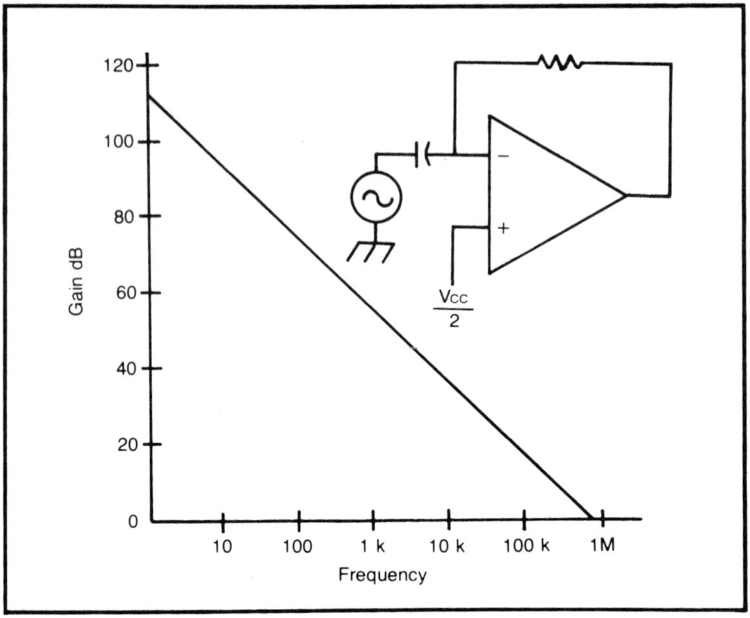

Fig. 5-13. Open-loop frequency response for the LM324 operational amplifier.

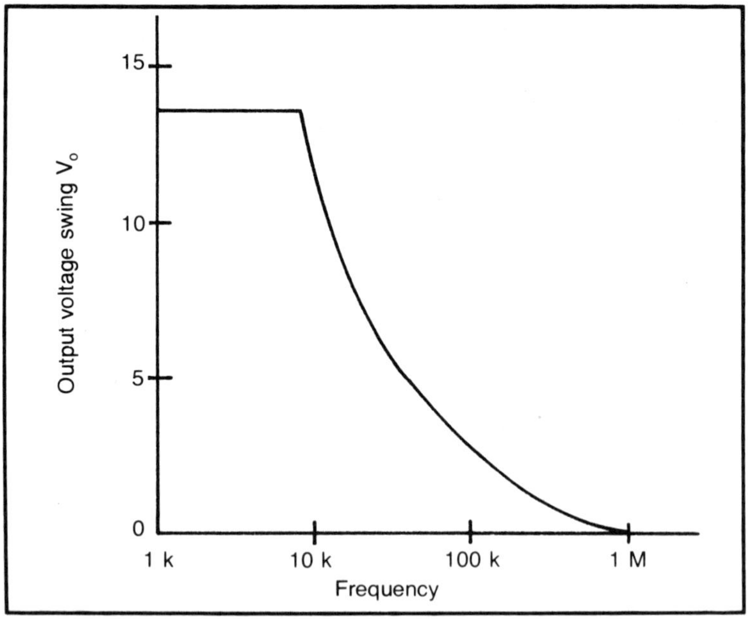

Fig. 5-14. Large signal frequency response for the LM324 operational amplifier showing limiting at the higher frequencies.

gain of 100, a 2 kΩ load, and a V+ of +15 volts. This curve shows the affect of frequency on the maximum output voltage swing. This is basically the slew rate of the amplifier, which is the maximum change of voltage per unit time. The flat portion of the curve is the limiting due to the dc characteristic. The slew rate is the curved portion of the curve, and works out to be about 0.1 volt/microsecond. This is the main limiting factor for high frequency operation. Trying to operate at a higher frequency will reduce the output amplitude as the frequency is increased. Between 500 kHz and 1 MHz, the output will be reduced to zero.

MC1306 Power Amplifier

The MC1306 is a complementary power amplifier and preamplifier designed to deliver ½ watt into an 8 ohm speaker with a 3 mV input into the preamplifier and 12 volts dc supply voltage. The gain and bandwidth are adjustable, and the IC has low zero signal current drain.

Table 5-15 gives the absolute maximum ratings. This device is operated from a single power supply, typically 12 volts dc. With a maximum V+ of 15 volts, the 12 volt supply is about as high as possible and still be safe. The maximum power dissipation of 625 milliwatts is in excess of the half watt output, giving a margin of safety.

The electrical characteristics are given in Table 5-16. The open-loop gains of the preamplifier and amplifier are both given in volts output per volts input. The open-loop gains are established using feedback networks in a manner similar to that used for operational amplifiers.

Sensitivity is the lowest signal which can be recognized by the preamplifier. Signals below this level are either lost or distorted badly. The 3 mV sensitivity is within the capability of low level devices such as phonograph pickup cartridges, microphones, and tape recorder heads. The signal-to-noise ratio is a measure of the

Table 5-15. MC1306 Absolute Maximum Ratings.

Rating	Symbol	Value	Unit
Power Supply Voltage	V^+	15	Volts
Load Current	I_L	400	mA
Power Dissipation	P_d	625	mW
Operating Temperature	T_A	0 to +75	°C
Storage Temperature	T_{stg}	−65 to +150	°C

Table 5-16. MC1306 Power Amplifier Electrical Characteristics

Parameter	Symbol	Typ	Unit
Open-Loop Voltage Gain Preamp R_L = 1 K Power amp R_L = 16 ohms	A_{VOL}	 270 360	V/V
Sensitivity	S	3	mV
Signal-to-Noise Ratio	S/N	55	dB
Total Harmonic Distortion	THD	0.5	%
Quiescent Output Voltage	V_o	$V^+/2$	Volts
Output Power	P_o	570	mW
Current drain (zero input)	I_D	4	mA
Power Dissipation (zero input)	P_d	36	mW

ability of the circuit to amplify the signal over the noise, both internally and externally generated. The larger this number the less electrical noise that will appear at the output as signal distortion. The total harmonic distortion is the distortion of the signal caused by the circuit components. This distortion, if excessive, can be heard in the output. The quiescent output voltage is the output with no input signal. Since this is a single power supply device operating in the class A mode, this voltage should be ½ of the power supply voltage. This places the operating point at the midpoint of the linear portion of the response curves.

Figure 5-15 shows the block diagram and the pin numbers. This device comes in an 8 pin plastic DIP package. Both the preamplifier and the power amplifier have single input and output pins. The power amplifier has an output ground pin. This pin is connected to the emitter of the output transistor. Normally this pin is connected to circuit ground.

Figure 5-16 shows the efficiency of the power amplifier for different power supply voltages and load resistance. The efficiency peaks out to about 50 percent for an 8 ohm load. This is good for a class A amplifier where there is always some dc current flow through the output, even for zero signal input.

Figure 5-17 shows the output power for various power supply voltages. Curves are given for loads of 8 ohms and 16 ohms.

ADVANCED 7400 DEVICES

Next, let us look at the specifications for some advanced 7400 series

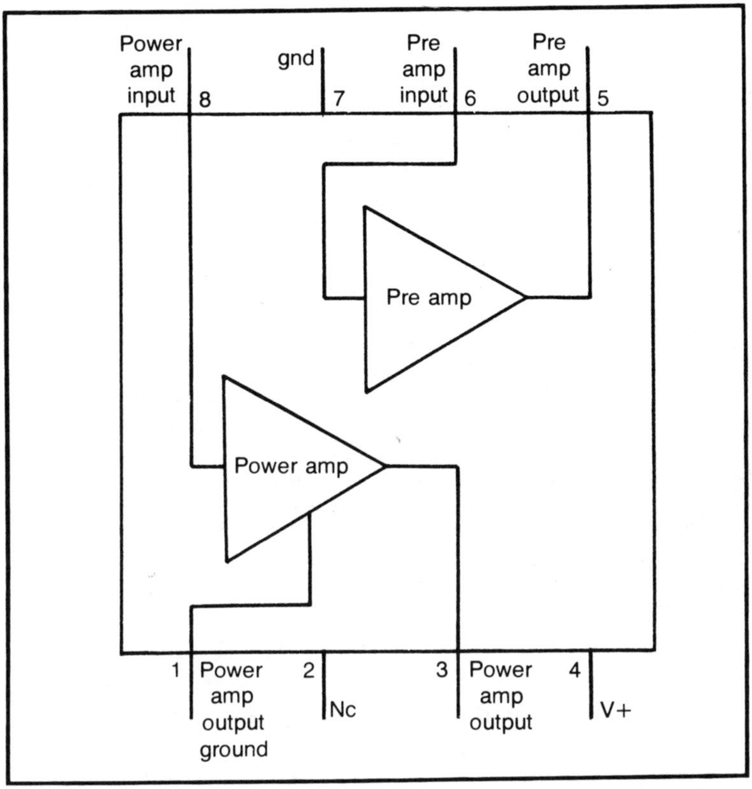

Fig. 5-15. Functional drawing and pin definition for the MC1306 power amplifier.

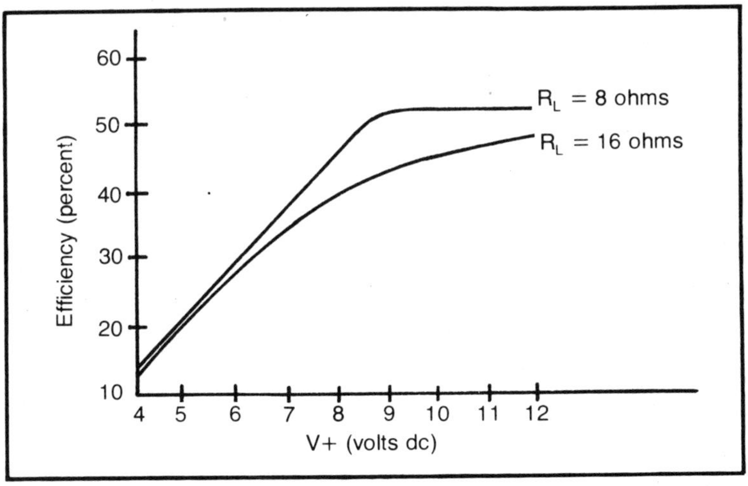

Fig. 5-16. Efficiency vs power supply voltage for the MC1306 power amplifier.

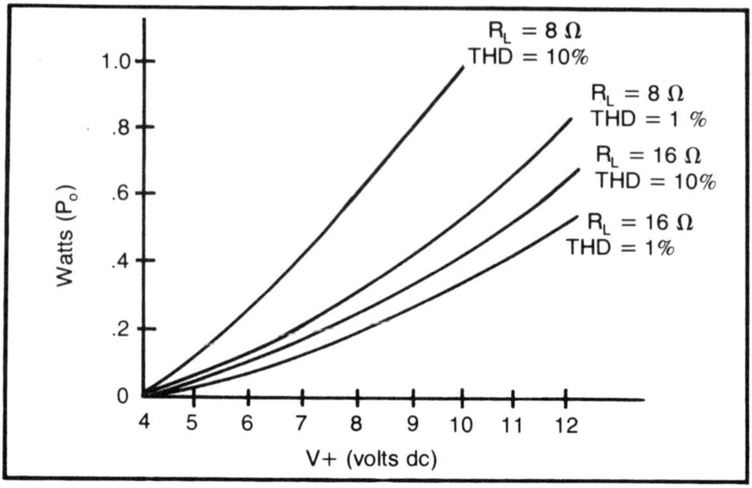

Fig. 5-17. Output power vs power supply voltage for the MC1306 power amplifier.

TTL integrated circuits. These are more complex than the simple gates covered in the beginning of this chapter. These complex ICs employ several gates into each of the logic circuits. The 7447, 7474, 7485, and 7490, have their own set of specifications because it takes different parameters and values to define the function of each IC. As previously mentioned, the absolute maximum ratings for the 7400 series ICs apply to these ICs. The specifications for these integrated circuits include the following:

Input and output loading
Dc characteristics
Ac characteristics
Pin configuration
Logic diagram
Any applicable truth tables
Any required waveshapes
A functional introduction

7447 BCD-to-Seven-Segment Decoder/Driver

This integrated circuit is used to drive a seven-segment LED display from a 4-bit BCD input. A BCD number is the binary equivalent of the decimal digit. Table 5-17 gives the BCD numbers in binary for the decimal numbers 0 through 9. This comprises one

BCD	Binary
0	0000
1	0001
2	0010
3	0011
4	0100
5	0101
6	0110
7	0111
8	1000
9	1001

Table 5-17. The Binary Equivalents for Decimal Numbers 0 to 9

digit, and the numbers are repeated for the next higher or lower digit.

Binary numbers are a method of counting using only two numbers, zero and one, for each digit. Thus, digital circuits can be

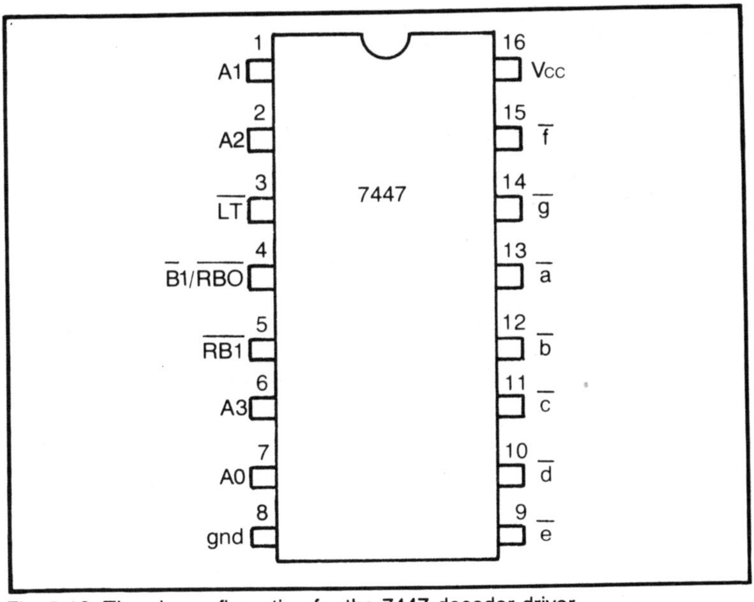

Fig. 5-18. The pin configuration for the 7447 decoder driver.

used because they have only two states, zero and one, high and low, etc. Referring to Table 5-17, the binary equivalents of the numbers are given. Using only 0 and 1 to represent each digit, the binary numbers count up in sequence for decimal 0 through 9. This is called BCD, or binary coded decimal, numbers because the binary equivalent of each decimal digit is represented. Notice that there are some of the binary number combinations that do not have BCD equivalents. These numbers, binary 1010 to binary 1111 are not used in BCD.

Figure 5-18 shows the pin configuration for the 7447 decoder/driver IC. Notice that this is a 16-pin DIP package. It contains about 32 multiple input gates of different types and 11 inverters. This IC decodes the BCD inputs, A_0, A_1, A_2, and A_3 and activates the required segment outputs, a, b, c, d, e, f, and g, so that the

Table 5-18. Functional Pin Description for the 7447

Pin	Symbol	Definition
1	A1	BCD input, next to the lowest order
2	A2	BCD input, next to the highest order
3	$\overline{LT}$	Light test input
4	$\overline{BI}/\overline{RBO}$	Ripple blanking output
5	$\overline{RBI}$	Ripple blanking input
6	A3	BCD input, high order
7	A0	BCD input, low order
8	GND	Power supply ground
9	$\overline{e}$	Segment e output
10	$\overline{d}$	Segment d output
11	$\overline{c}$	Segment c output
12	$\overline{b}$	Segment b output
13	$\overline{a}$	Segment a output
14	$\overline{g}$	Segment g output
15	$\overline{f}$	Segment f output
16	Vcc	Power supply voltage

Table 5-19. 7447 Decoder/Driver Input Loading and Output Drive Table

Pins	High State	Low State
A0-A3	40 μA	−1.6 mA
$\overline{LT}$	40 μA	−1.6 mA
$\overline{a} - \overline{g}$	250 μA	40 mA

equivalent decimal number is formed on the seven-segment display. Table 5-18 gives the functional pin description for this IC. The $\overline{LT}$ input is a lamp test input, used to test all the segments of the display with a single input. Many display panels have a lamp test switch which illuminates all the panel lights at the same time. The $\overline{LT}$ signal is active low, that is grounding this input will activate the lamp test, overriding any other input. The $\overline{RBI}$, or ripple blanking input is used to suppress leading or training zeros in a multidigit display. The $\overline{BI/RBO}$ pin is a dual function pin. It is the ripple blanking output and brightness input. For leading and trailing zero blanking, $\overline{RBO}$ is connected to the $\overline{RBI}$ of the next lower digit. The $\overline{BI}$ provides a means for modulating the intensity.

Table 5-19 gives the input loading and output drive capability for the 7447 decoder/driver. The current is given for both the high and low state of the given signal. Notice that the segment outputs will drive 40 mA. If all seven segments are on at once, this gives 280 mA drive possible through the chip. Normally, the output is current limited by series resistors to 5 to 16 mA, depending on the seven-segment display used, and the brightness required. There are some special displays which use higher currents, but they will not be covered here.

The outputs of this IC are the open collector type, so that the output high voltage is that voltage which is applied to the output by the load circuit. From the output drive chart, it is apparent that the

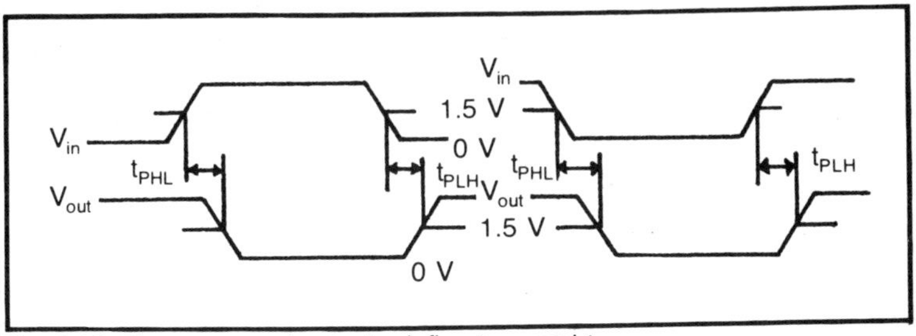

Fig. 5-19. The ac waveshapes to define t_{PHL} and t_{PLH}.

Table 5-20. The DC Characteristics for the 7447 Decoder

Parameter		Value
V_{OH}	Maximum voltage applied to segment outputs	15 volts
I_{OH}	Output high current at the segment outputs	250 μA
V_{OL}	Low output voltage at the segments (I_{OL} = 8 mA)	0.4 volts
I_{OL}	Current sink capability of the segment outputs.	40 ms
Icc	Supply current	103 mA

segment outputs are active low. The workings and use of seven-segment displays is covered in more detail in Chapter 6.

Table 5-20 gives the dc characteristics for this IC. The output high voltage, or the voltage applied to the segment outputs by external circuitry is 15 volts. Figure 5-19 shows the ac waveshapes which define the propagation delay. This value is 100 ns for both t_{PHL} and t_{PLH}.

Table 5-21 gives the truth table for the 7447 decoder/driver. The inputs correspond to the BCD inputs and the outputs are the segment outputs (active low). The BCD inputs are active high while the segment outputs are active low. These correspond to the segments of the seven-segment display, and will become clearer when the section on displays in Chapter 6 is studied.

Table 5-21. Input Versus Output Truth Table for the 7447

Decimal Number	Inputs				Outputs						
	A3	A2	A1	A0	$\bar{a}$	$\bar{b}$	$\bar{c}$	$\bar{d}$	$\bar{e}$	$\bar{f}$	$\bar{g}$
0	L	L	L	L	L	L	L	L	L	L	H
1	L	L	L	H	H	L	L	H	H	H	H
2	L	L	H	L	L	L	H	L	L	H	L
3	L	L	H	H	L	L	L	L	H	H	L
4	L	H	L	L	H	L	L	H	H	L	L
5	L	H	L	H	L	H	L	L	H	L	L
6	L	H	H	L	H	H	L	L	L	L	L
7	L	H	H	H	L	L	L	H	H	H	H
8	H	L	L	L	L	L	L	L	L	L	L
9	H	L	L	H	L	L	L	H	H	L	L

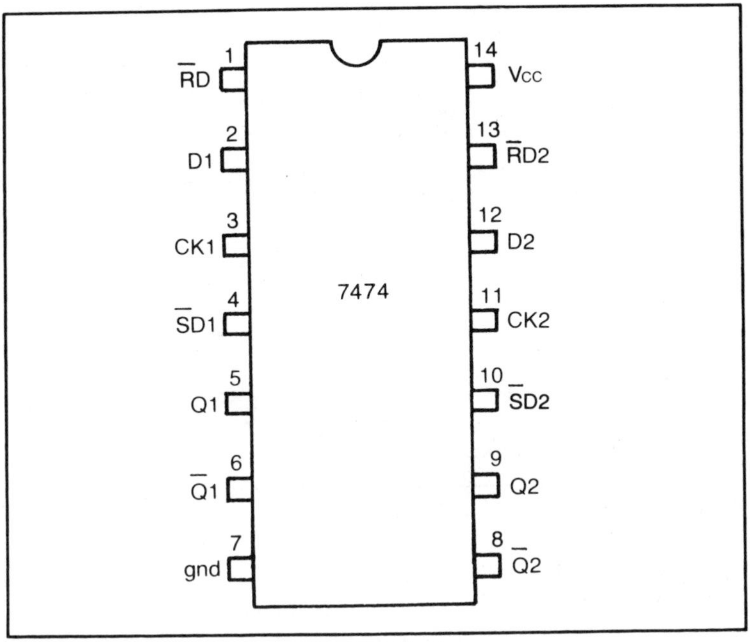

Fig. 5-20. Pin configuration for the 7474 flip-flop.

7474 Dual D-Type Flip-Flop

A flip-flop is a circuit which, when triggered to one state, will remain in that state until the conditions are met to trigger the circuit to the opposite side. They have two outputs, Q and $\overline{Q}$, which are complements of each other. When one is in the high state the other is low. The flip-flop will retain its previous state regardless of the input conditions as long as the trigger conditions are not met. This is one of the basic uses of the flip-flop, where it acts as a *memory*, and remembers the condition for use in following time periods.

Another widespread use of flip-flops is in countdown circuits, where the output pulse rate is a submultiple of the input pulse rate. This type of application is found in timing circuits, and for the generation of timing pulses. There are many more applications, such as shift registers and multiplexers. The flip-flop is one of the basic circuits used in computers.

Figure 5-20 shows the pin configuration for the 7474 IC. Figure 5-21 shows the logic representation for this flip-flop. This is the standard D-type flip-flop symbol as used in more complex logic diagrams. The functional description of the pins is given in Table 5-22. Notice that this is a dual IC with two identical and independent

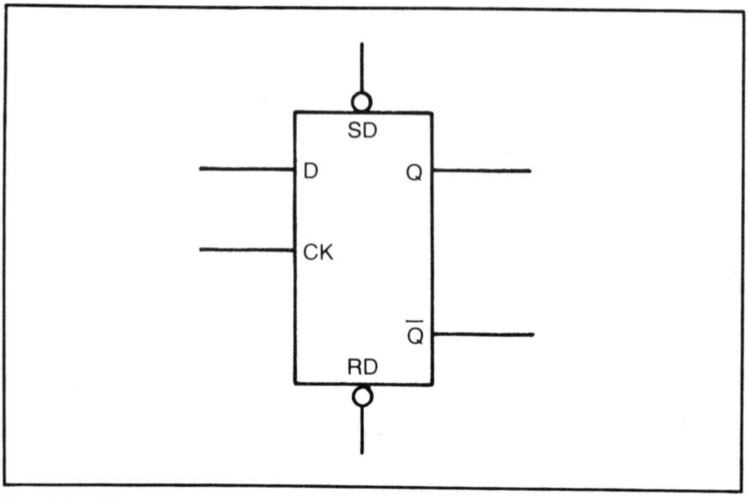

Fig. 5-21. The logic symbol for the 7474 flip-flop.

flip-flops. The pin symbols for one flip-flop have the subscript 1, while the other has the subscript 2. Table 5-22 gives the pin numbers for flip-flop number 1 first.

Table 5-23 is the mode select table, and gives the response for the various inputs. Refer to this table in the following discussion. The set and reset inputs are asynchronous active low inputs. Anytime either of these inputs are active, no matter what the condition of the other inputs, the flip-flop is either set or reset. The set condition is with the Q output high, and the reset condition is with the Q output low. Remember that the $\overline{Q}$ output is always the opposite of the Q output, so $\overline{Q}$ will be low in the set condition and high in the reset condition. The set and reset inputs allow setting the flip-flop to a predetermined state.

The D input is the data input. The data at the D input is loaded into the flip-flop and appears at the Q output on the low to high transition of the clock.

Pins	Symbol	Description
1, 13	RD	Reset input
2, 12	D	Data input
3, 11	CK	Clock input
4, 10	SD	Set input
5, 9	Q	Q output
6, 8	$\overline{Q}$	Inverted output

Table 5-22. Functional Pin Definition for the 7474 Flip-Flop

Table 5-23. The Mode Select Table for the 7474

Operating mode	Inputs				Outputs	
	$\overline{SD}$	$\overline{RD}$	CK	D	Q	$\overline{Q}$
Asynchronous Set	L	H	X	X	H	L
Asynchronous Reset	H	L	X	X	L	H
Undetermined	L	L	X	X	H	H
Load 1 (set)	H	H	1/h	h	H	L
Load 0 (reset)	H	H	1/h	1	L	H

X don't care
1/h low to high clock transition
h High level one set up time prior to clock
1 Low level one set up time prior to clock

Figure 5-22 shows the logic diagram for the 7474 flip-flop. It is comprised of six NAND gates for each of the two flip-flops. As an exercise in logic circuits, the reader can analyze the circuits by setting up a truth table and tracing through the operation for all conditions of the inputs.

Table 5-24 gives the input loading and the output drive for the 7474 IC. The input loads are the drive required to assure reliable operation of that input. Notice that the output can drive a 16 mA active low load, but only a 400 μA active high load. This is typical of TTL circuits in that they are typically active low.

The only dc characteristic given is the maximum dc supply current, which is 30 mA worst case. This means that the most current that the chip can pull from the power supply is 30 mA. But, from the loading and drive table (Table 5-24) the maximum output current for the Q and $\overline{Q}$ outputs is 16 mA in the low state, giving 32 mA for both flip-flops in the IC. This is possible because this is the low output current, and the current is supplied by the load, not by the IC. The output circuitry of the IC routes the current to ground. So, the IC is not drawing this low state current from the power supply.

The ac characteristics for the 7474 IC are given in Table 5-25. The waveshape numbers refer to the waveshapes given in Fig. 5-23. The minimum clock frequency is 15 MHz. This is the highest frequency that any 7474 IC chip will operate at reliably. The remaining values are the propagation delays between waveshapes.

Table 5-26 gives the ac set up requirements. Again, the waveshape numbers refer to the waveshapes given in Fig. 5-23. The set up requirements establish the timing requirements for reliable operation. Not meeting the minimum requirements can

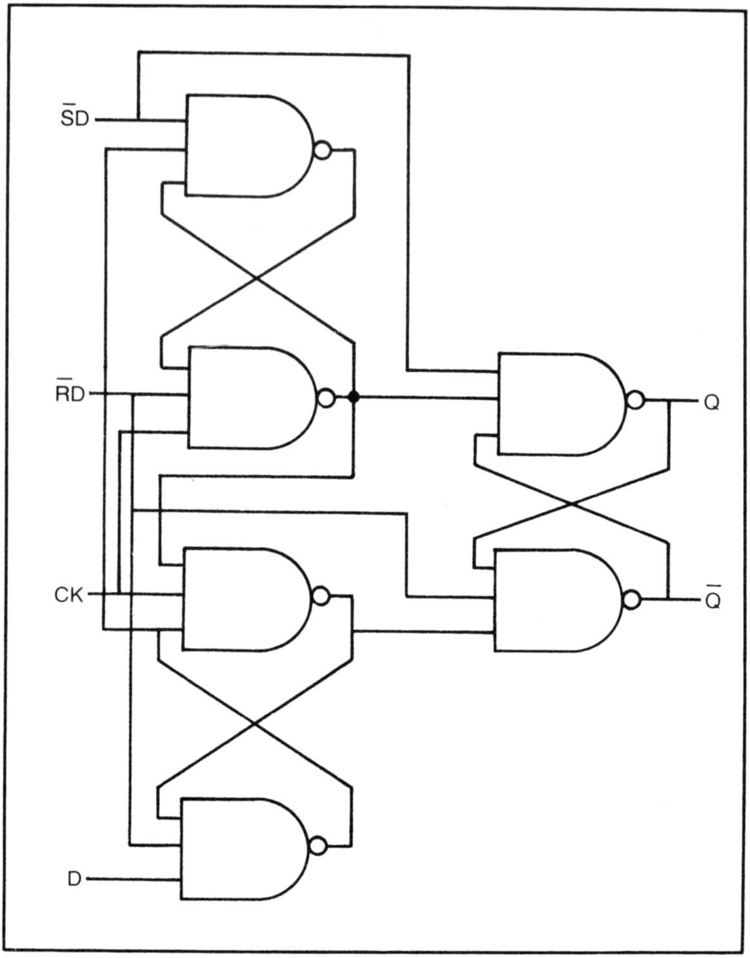

Fig. 5-22. Logic diagram for the 7474 flip-flop showing the equivalent internal gates.

Table 5-24. Input Loading and Output Drive for the 7447

Pins	High State	Low State
D Data Input	40 μA	−1.6 mA
CK Clock Input	80 μA	−3.2 mA
$\overline{SD}$ Set Input	80 μA	−1.6 mA
$\overline{RD}$ Reset Input	120 μA	−3.2 mA
Q & $\overline{Q}$ Outputs	−400 μA	16 mA

Table 5-25. The AC Characteristics for the 7474 Flip-Flop

Parameter	Waveshape	Value
f_{max}—Max clock freq	1	15 MHz min
t_{PLH}—Propagation delay	1	25 ns max
t_{PHL}—Clock to output	1	40 ns max
t_{PLH}—Set or reset to	2	25 ns max
t_{PHL}—Output	2 (CK high)	40 ns max
t_{PHL}—Set or reset to output	2 (CK low)	40 ns max

cause operational problems when using this IC.

The use of the 7474 flip-flop and all the advanced 7400 series ICs are covered in Chapter 6 and in the projects given in Chapter 8.

7485 4-Bit Magnitude Comparator

The 7485 is a 4-bit magnitude comparator which compares two 4-bit

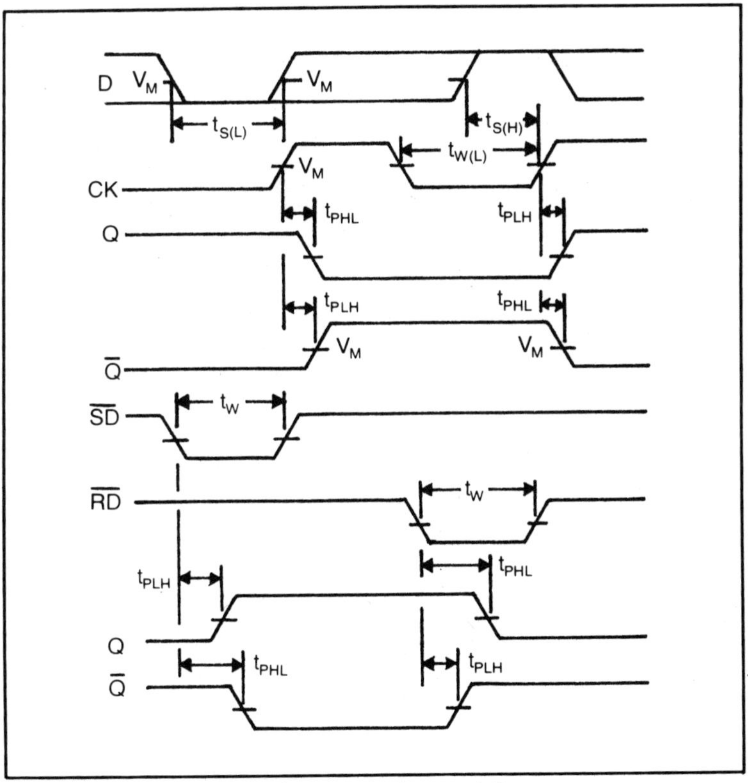

Fig. 5-23. Waveshapes to define the ac characteristics.

Table 5-26. Ac Set Up Requirements for the 7474 Flip-Flop

	Parameter	Value	Waveshape
$t_{w(H)}$	Clock pulse width (high)	30 ns min	1
$t_{w(L)}$	Clock pulse width (low)	37 ns min	1
$t_{w(L)}$	Set or reset pulse width (low)	30 ns min	2
$t_{s(H)}$	Setup time (high) data to clock	20 ns min	1
$t_{s(L)}$	Setup time (low) data to clock	20 ns min	1
t_h	Hold time data to clock	5 ns min	1

binary or BCD numbers, and gives three possible outputs. The possible outputs are:

word A greater than word B
word A equal to word B
word A less than word B

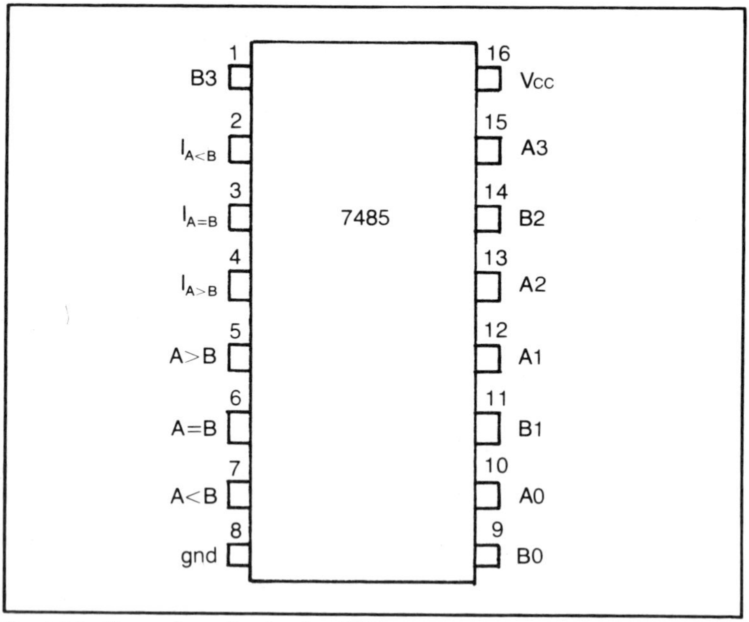

Fig. 5-24. Pin configuration for the 7485 comparator.

The individual bits of the 4-bit binary word are compared one bit at a time, starting with the high order bits and proceeding to the low order, until the first noncompare or until all bits have been compared. There are cascading inputs to allow the comparator to be expanded to any word size. This IC contains 30 logic gates and is quite complex; therefore, the logic diagram is not shown.

Figure 5-24 gives the pin configuration for the 7485 and Fig. 5-25 gives the logic symbol. This is a 16-pin DIP package with Vcc at pin 16 and ground at pin 8. The functional definition of the pins is given in Table 5-27. The inputs for word A are A_0, A_1, A_2, and A_3 with A_3 being the high order bit. The inputs for word B are B_0, B_1, B_2, and B_3.

As an example of the compare operation, let me give some examples.

word A = 1001
word B = 1001

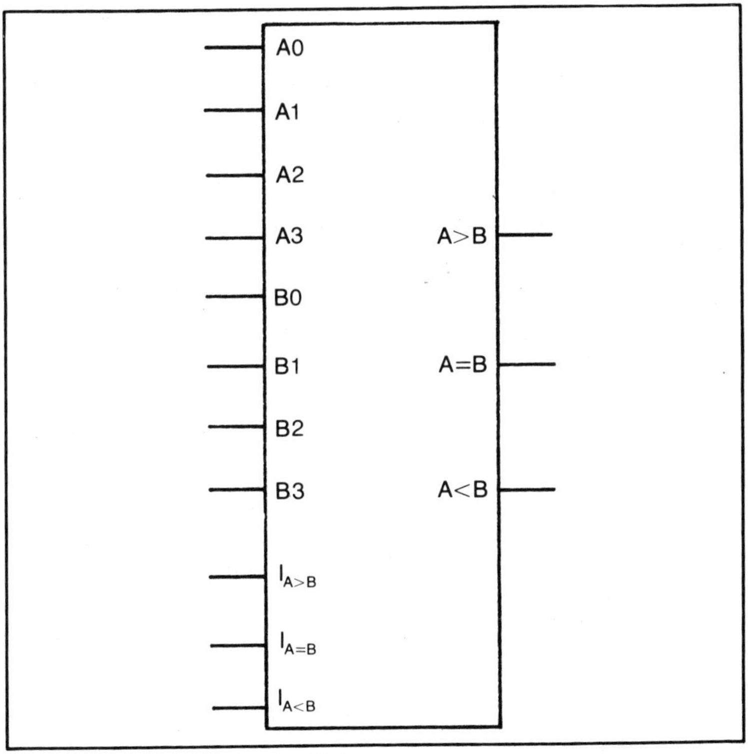

Fig. 5-25. Logic symbol for the 7485 comparator.

Table 5-27. Definition of the Pins for the 7485 Comparator

Pin	Symbol	Definition
1	B3	High order bit input for word B
2	$I_{A<B}$	Cascade input, A less than B
3	$I_{A=B}$	Cascade input, A equal B
4	$I_{A>B}$	Cascade input, A greater than B
5	A>B	Output, A greater than B
6	A=B	Output, A equal B
7	A<B	Output, A less than B
8	Gnd	Ground
9	B0	Low order input, word B
10	A0	Low order input, word A
11	B1	Next to low order input, word B
12	A1	Next to low order input, word A
13	A2	Next to high order input, word A
14	B2	Next to high order input, word B
15	A3	High order input, word A
16	Vcc	Supply Voltage

The output for A = B will be active (high) because the bits are identical. In the next example the words are different.

word A = 1000
word B = 0111

The output for A greater than B will be active because word A is larger than word B. The comparison starts with the high order bits. If they are equal, the next lower bits are compared. This continues until all the bits are compared, or a noncompare is found. In this case, a noncompare (A greater than B) was found on the first compare. Try one more example.

word A = 0010
word B = 0011

The output for A less than B will be active in this case.

Table 5-28 gives the truth table for the 7485 comparator. An X in any of the input condition columns means that the circuit doesn't care what that input is for the conditions on that line because the results were obtained at a higher order bit. This table gives the outputs for the various conditions of the inputs. If the cascade inputs are not in use, then they must simulate the $I_{A=B}$ condition so that the condition is determined by the inputs A and B, and not the cascade inputs. For this condition, input A greater than B (pin 4) and input A less than B (pin 2) should be grounded and input A equals B should

Table 5-28. Truth Table for the 7485 Comparator

Comparing Inputs				Cascade			Outputs		
A3,B3	A2,B2	A1,B1	A0,B0	$I_{A>B}$	$I_{A<B}$	$I_{A=B}$	A>B	A<B	A=B
A3>B3	X	X	X	X	X	X	H	L	L
A3<B3	X	X	X	X	X	X	L	H	L
A3=B3	A2>B2	X	X	X	X	X	H	L	L
A3=B3	A2<B2	X	X	X	X	X	L	H	L
A3=B3	A2=B2	A1>B1	X	X	X	X	H	L	L
A3=B3	A2=B2	A1<B1	X	X	X	X	L	H	L
A3=B3	A2=B2	A1=B1	A0>B0	X	X	X	H	L	L
A3=B3	A2=B2	A1=B1	A0<B0	X	X	X	L	H	L
A3=B3	A2=B2	A1=B1	A0=B0	Not connected			L	L	H
A3=B3	A2=B2	A1=B1	A0=B0	H	L	L	H	L	L
A3=B3	A2=B2	A1=B1	A0=B0	L	H	L	L	H	L
A3=B3	A2=B2	A1=B1	A0=B0	L	L	H	L	L	H

be tied high. This will assure proper operation. Note from the table that the active condition for the outputs is high.

The truth table can be used to work out the above examples by starting at the first line and continuing down until the numbers match the conditions of the line. Remember that the most significant bit is A_3 and B_3 and is the left digit on the binary numbers.

The input loading and output drive table for the 7485 comparator is given in Table 5-29. Being a TTL type IC, the low condition is the driven condition and requires the most current. Notice that the input drive currents are three times that required for a normal, simple TTL gate. This is because of the large number of gates inside the IC, and the number of gates driven by each input. The drive capability for the outputs is about standard for TTL logic circuits. The maximum supply current drawn by the 7485 is 88 mA.

Table 5-30 gives the ac characteristics for this IC. These are propagation delay times. Refer to Fig. 5-19 for the waveshapes defining the parameters.

Pins	High State	Low State
A0–A3	120 μA	−4.8 mA
B0–B3	120 μA	−4.8 mA
$I_{A=B}$	120 μA	−4.8 mA
$I_{A>B}$	40 μA	−1.6 mA
$I_{A<B}$	40 μA	−1.6 mA
A=B	−400 μA	16 mA
A>B	−400 μA	16 mA
A<B	−400 μA	16 mA

Table 5-29. Input Loading and Output Drive for the 7485

Table 5-30. 7485 Comparator Ac Characteristics

Parameter		Waveshape	Value
t_{PLH} t_{PHL}	Propagation delay A or B input to A<B, A>B output	1	26 ns max 30 ns max
t_{PLH} t_{PHL}	Propagation delay A or B input to A = B	2	35 ns max 30 ns max
t_{PLH} t_{PHL}	Propagation delay exp inputs to A<B, A>B	1	11 ns max 17 ns max
t_{PLH} t_{PHL}	Propagation delay $I_{A=B}$ input to A = B	2	20 ns max 17 ns max

7490 Decade Counter

The 7490 decade counter is a 4-bit binary counter connected to count only from 0 to 9 then reset to 0. The input is a serial pulse train and the output is four parallel pulses which equals the BCD equivalent of the input count. The counter contains four flip-flops. One is connected as a divide-by-two counter and the other three are connected as divide-by-5 counters. The two counter sections are not connected, and the output of the first stage must be connected to the input of the second stage when using the 7490 as a decade counter. Two set and reset inputs are provided. These are active high inputs, and at least one set and one reset must be low for the circuit to operate.

Table 5-31. Functional Pin Definition for the 7490 Counter

Pin	Symbol	Definition
1	$\overline{CK}1$	Divide-by-5 section clock input
2	MR1	Master reset input #1
3	MR2	Master reset input #2
4	NC	No connection
5	Vcc	Supply voltage
6	MS1	Master set input #1
7	MS2	Master set input #2
8	Q2	Next to high order BCD output
9	Q1	Next to low order BCD output
10	Gnd	Ground
11	Q3	High order BCD output
12	Q0	Low order BCD output
13	NC	No connection
14	$\overline{CK}0$	Divide-by-2 section clock input

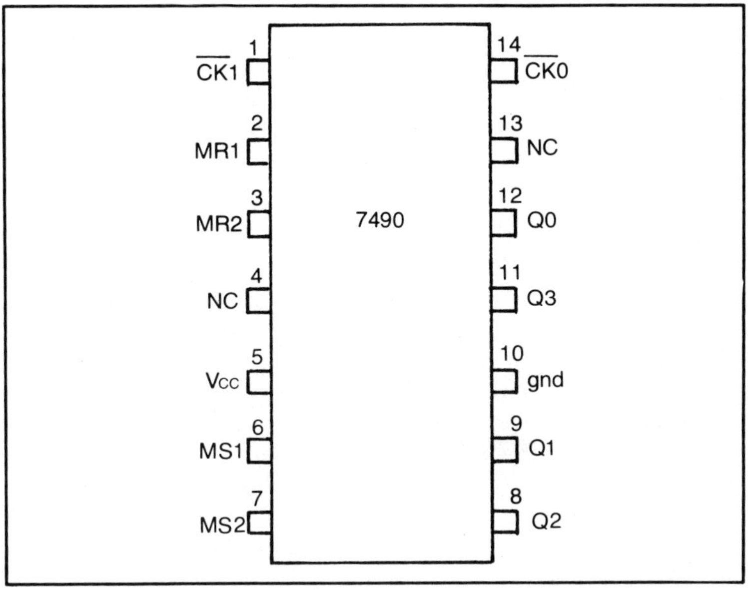

Fig. 5-26. Pin configuration for the 7490 decade counter.

Figure 5-26 gives the pin configuration for the 7490. The function of the individual pins is given in Table 5-31. Figure 5-27 is the logic symbol and Fig. 5-28 is the logic diagram for this IC. Q_0 is the output of the first stage, the divide by two counter, and $\overline{CK_1}$ is

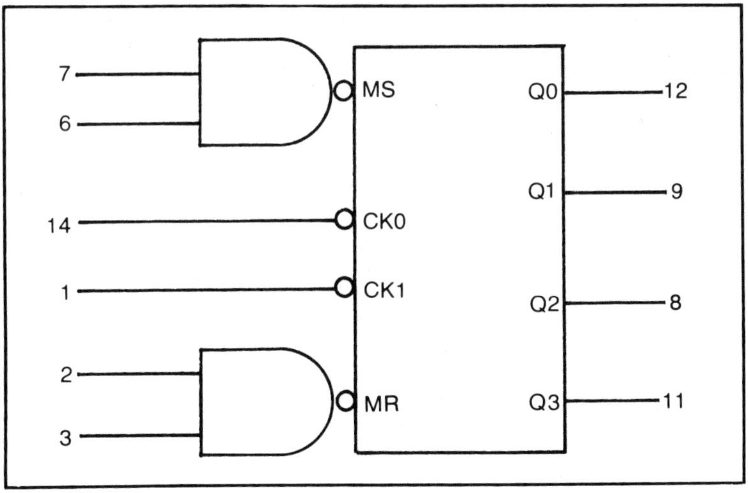

Fig. 5-27. The logic symbol for the 7490 counter.

Pin	High State	Low State
$\overline{CK}0$	80 μA	−3.2 mA
$\overline{CK}1$	160 μA	−6.4 mA
MR1. MR2	40 μA	−1.6 mA
MS1. MS2	40 μA	−1.6 mA
Q0 thru Q3	−800 μA	16 mA

Table 5-32. Input Loading and Output Drive for the 7490 Counter

the input for the divide by 5 section. These two must be connected together externally. In this condition, $\overline{CK}_0$ is the serial input and Q_0, Q_1, Q_2, and Q_3 are the parallel BCD outputs, with Q_3 being the high order bit. MS_1, MS_2, MR_1, and MR_2 are the master set (MS) and master reset (MR) inputs. These inputs give the capability of presetting the counter to 0 (reset) or 9 (set).

Table 5-32 is the input load and output drive table. Notice that the set and reset inputs require one unit load drive, as with typical TTL ICs, and can be driven from a simple gate. But $\overline{CK}_0$ and $\overline{CK}_1$ require more drive. From the logic diagram of Fig. 5-28, $\overline{CK}_1$ drives two flip-flops and $\overline{CK}_0$ drives one. Referring back to the 7474 flip-flop input loading table, the clock input requires 2 unit loads, or 3.2 mA. This is where the 3.2 mA drive for $\overline{CK}_0$ and 6.4 mA drive for $\overline{CK}_1$ comes from.

Table 5-33 is the mode selection truth table. This table gives the condition of the outputs for the set and reset inputs. The 7490 will count when one or more set inputs are low and one or more reset inputs are low. The reset condition, when all outputs are low, is achieved when both reset inputs are high and one set input is low. The set condition, when Q_0 and Q_3 are high, is achieved when both set inputs are high, and one reset input is low. The output conditions when all four set and reset inputs are high is not defined, and should not be allowed.

Table 5-33. Mode Selection Truth Table for the 7490

Reset/Set Inputs				Outputs			
MR1	MR2	MS1	MS2	Q0	Q1	Q2	Q3
H	H	L	X	L	L	L	L
H	H	X	L	L	L	L	L
X	X	H	H	H	L	L	H
L	X	L	X	Count			
X	L	X	L	Count			
L	X	X	L	Count			
X	L	L	X	Count			

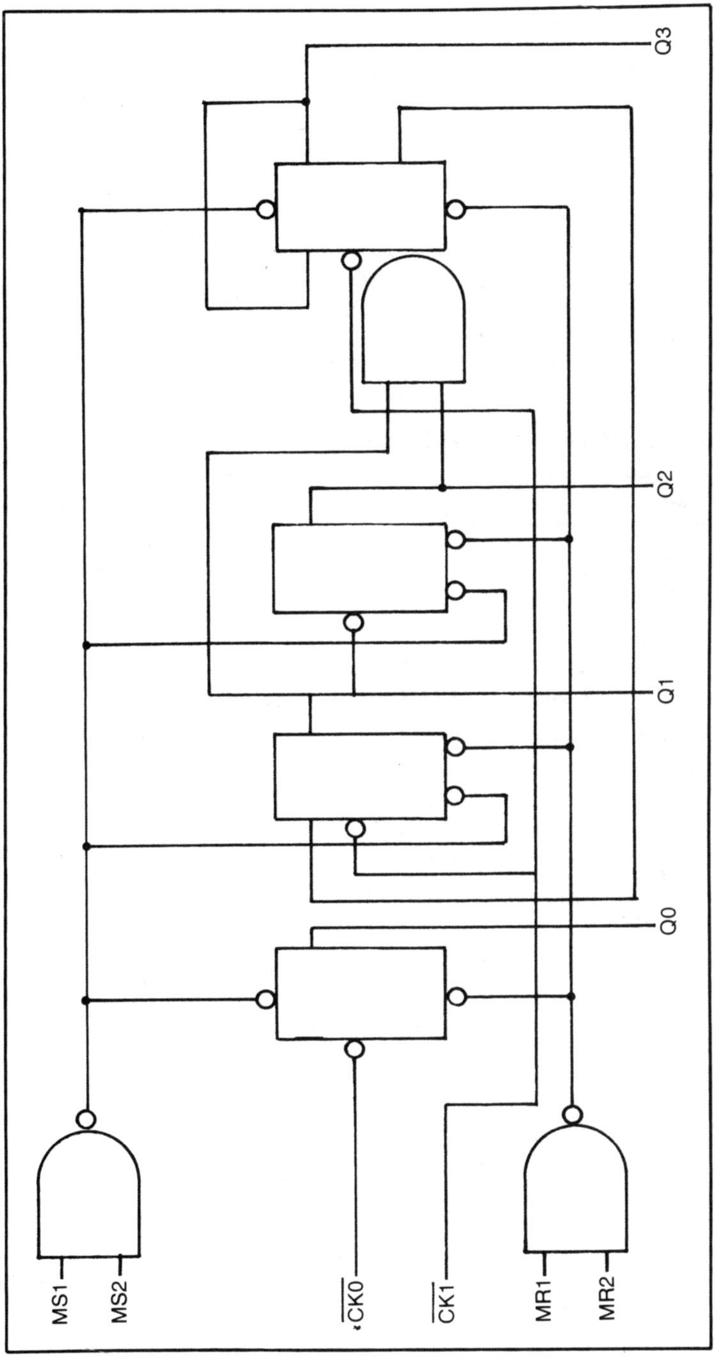

Fig. 5-28. The logic diagram for the 7490 counter.

Count	Output			
	Q0	Q1	Q2	Q3
0	L	L	L	L
1	H	L	L	L
2	L	H	L	L
3	H	H	L	L
4	L	L	H	L
5	H	L	H	L
6	L	H	H	L
7	H	H	H	L
8	L	L	L	H
9	H	L	L	H

Table 5-34. The BCD Count Sequence Table Showing the Binary Equivalents for Decimal Numbers 0 Thru 9

Table 5-35. The Ac Characteristics for the 7490 Decade Counter

Parameter		Waveshape	Value
f_{MAX}	$\overline{CK0}$ input count frequency	1	10 MHz min
f_{MAX}	$\overline{CK1}$ input count frequency	1	10 MHz min
t_{PLH} t_{PHL}	Propagation delay $\overline{CK0}$ to Q3 output	1	100 ns max

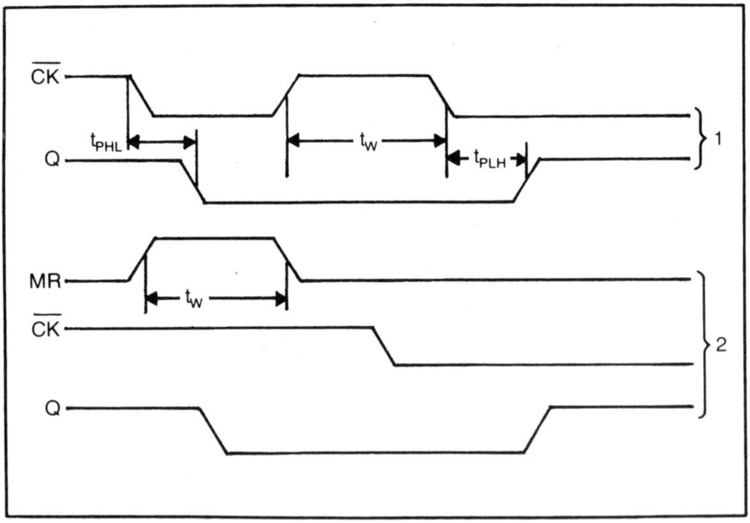

Fig. 5-29. Waveshapes for the 7490 decade counter.

Table 5-36. The Ac Set Up Requirements for the 7490 IC

Parameter	Waveshape	Min
t_w $\overline{CK0}$ Pulse Width	1	50 ns
t_w $\overline{CK1}$ Pulse Width	1	50 ns
t_w MR Pulse Width	2	50 ns

Table 5-34 is the BCD count sequence table. It gives the condition of the outputs, and thus the binary equivalent, for input counts of 0 thru 9. The next count after 9 will start the count over at 0, giving a total of 10 counts. Notice that Q_0 and Q_3 high (the set condition), is the decimal count of 9, so the next pulse after a set input will be count 0.

The only dc characteristic for this IC is the maximum supply current, which is 53 mA. This is the most current that the IC can draw from the power supply, and does not count current supplied by the drive or load circuits.

Table 5-35 gives the ac characteristics for the 7490 decade counter. The waveshapes refer to Fig. 5-29, and are mostly propagation delay times. The propagation delay is the time delay between one input and the resulting output, and is usually measured at 1.5 volts on the active going pulse.

The ac set up requirements for this IC are given in Table 5-36. Again, the waveshape numbers refer to those given in Fig. 5-29. All the parameters are defined on the waveshapes. The maximum clock frequency is 10 MHz as specified by the minimum value for the maximum count frequency. The minimum clock pulse width is 50 ns. The input requires a minimum of 50 ns pulse at not more than a 10 MHz pulse rate.

6

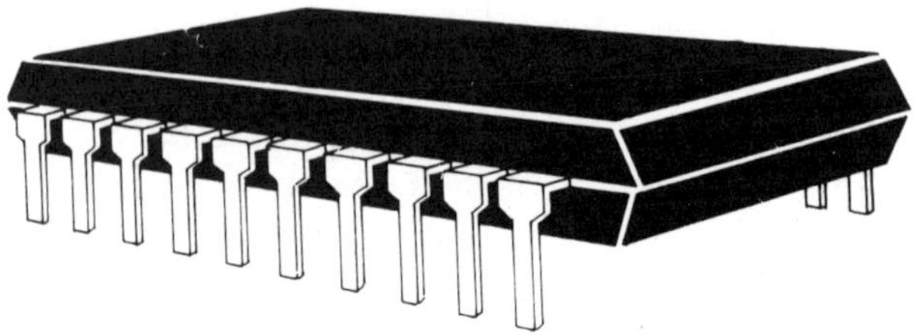

Digital Applications

In the past chapters we have discussed digital ICs and logic circuits in general. Seven ICs from the 7400 series of TTL digital ICs were discussed in detail.

In this chapter, I will present applications for these seven ICs. The application information will give you a better understanding of how digital ICs work and how to use them.

The construction of truth tables and the use of waveshapes for logic analysis is also discussed and demonstrated. One type of loading chart is presented and ac analysis discussed. These methods are used in the analysis of digital circuits. Any involved logic circuit must be analyzed to prevent possible timing, loading, or drive problems. These types of problems will only show up after the circuit has been in operation for a while, and then it may only show up occasionally.

CIRCUIT ANALYSIS

This section provides information on potential problems and details on circuit design and analysis. There are also some tips and tricks that you may find useful as you begin to build digital circuits. Many of the considerations presented here will help you build trouble-free projects.

Power Source

The power source used to supply the dc voltage for the ICs is one of the major considerations for circuits using several ICs. Since all the 7400 chips are powered by 5 volts, the regulation of the power supply must be sufficient to maintain the voltage within the limits. The dc voltage requirements, worst case, must be determined, and the power source must be capable of supplying this current. The worst case loading is determined by totaling the dc current requirements from the specification sheets.

In addition, power supply decoupling capacitors may be required if the power supply is located more than a few inches from the circuit. When the output of a TTL digital IC switches to a low output, and there is a load on the output, it is possible to generate spikes or glitches on the power supply line. This can interfere with the operation of other ICs that are switching at the same time. It can also result in excessive electrical noise being propagated throughout the system. One method used to overcome this is to use decoupling capacitors throughout the system.

Decoupling capacitors are capacitors placed across the ground and+5 volt bus lines at the IC, or at the approach to the integrated circuits. This is illustrated in Fig. 6-1. Typically a large and a small capacitor are used. The large capacitor is to store charge to smooth out the lower glitches and spikes. Since the larger capacitors, which

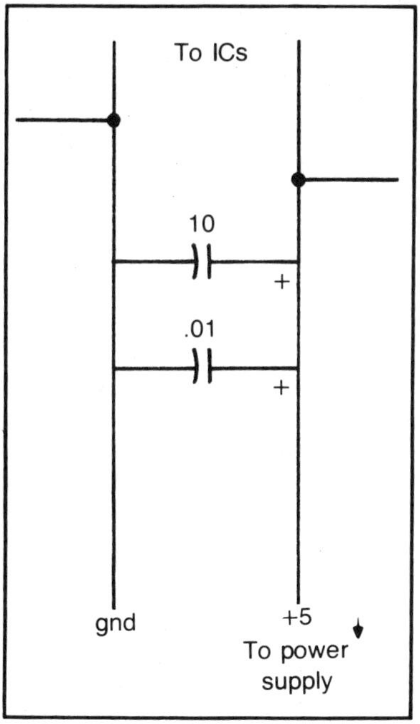

Fig. 6-1. Using decoupling capacitors across the voltage and ground buses.

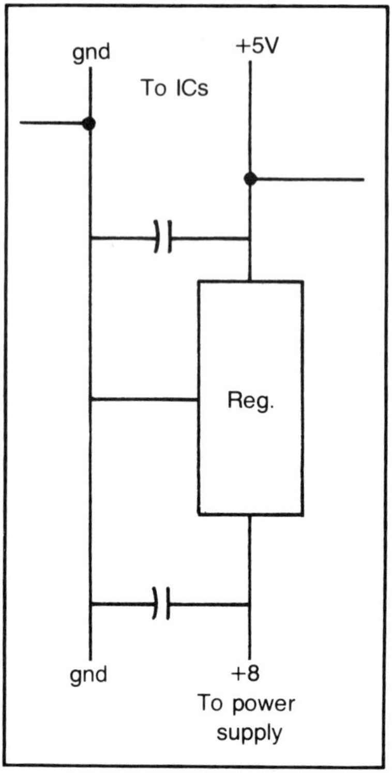

Fig. 6-2. Using separate three terminal regulators for separate groups of IC's to provide decoupling.

are physically small, have a high impedance at high frequencies, a small capacitor is also used. This smooths out the high frequency spikes. The values shown in the figure, 10 μF and 0.01 μF, are typical. The voltage rating should be at least twice the dc voltage, or at least 10 volts.

If there is a sensitive or critical circuit, the decoupling capacitors may be placed next to the IC, or even attached directly to the IC pins.

Another method commonly used to provide power supply decoupling is to provide separate voltage regulators for each group of circuits or each printed circuit board, as shown in Fig. 6-2. This requires that the main supply be 8 to 10 volts, but the regulation need not be as good. The separate regulators convert this to a well regulated 5 volts. With the advent of the three terminal regulators, this method is not expensive or difficult to use.

Fan-Out

The fan-out capability of a digital IC is the amount of circuits that the output can drive. It is usually given as unit loads, where one unit load is the typical drive requirement for a TTL IC, or 1.6 milliamp. If the unit load drive capability is not given, it can be calculated by dividing the maximum low output current from the specification sheet by 1.6 milliamp.

The loading is often checked for circuits with several integrated circuits by using a loading chart. The loading chart in Table 6-1, is for the circuit shown in Fig. 6-3. Each output has a separate line, and each input has one column. The outputs have their unit load capability listed, and the inputs have their drive requirements. So the total is the sum of the numbers on any one line. If the output is driving a circuit which is not an IC, but takes drive current, this load must be determined. When the current is found, the unit load is calculated by dividing by 1.6 milliamp. For instance, a circuit requiring 7.4 milliamps is 4 unit loads. The load must be determined for the 0 and 1 states.

Although Table 6-1 only shows the loading for the 0 state, a similar table may be required if there is 1 state loading. This is done in a similar manner, but may be carried through in milliamps in place

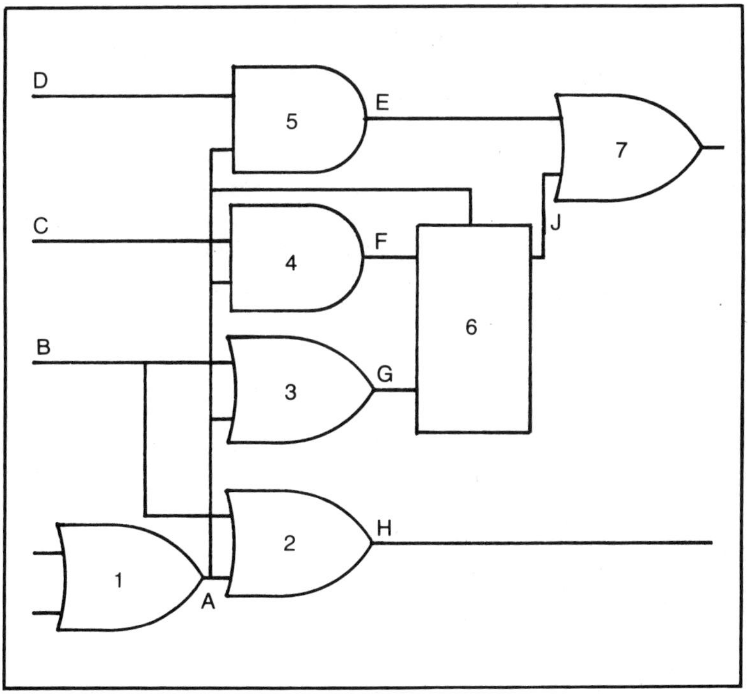

Fig. 6-3. Logic Circuit used for Tables 6-1 and 6-2.

Table 6-1. Output Loading Chart for Fig. 6-3

Signal	Integrated Circuit							Total
	1	2	3	4	5	6	7	
A 10		1	1	1	1	1.5		5.5
B 10		1	1					2
C 10				1				1
D 10					1			1
E 10							1	1
F 10						1.5		1.5
G 10						1.5		1.5
H 10								
J 3							1	1

of unit loads. This is because the drive capability of the 1 state for a typical digital IC is less than one unit load.

Table 6-2 is the power requirements chart for the circuit shown in Fig. 6-3. This gives the maximum current required to power the circuit.

Race Condition

One problem that can occur in TTL logic circuits is called a *race condition.* The typical race condition is where a short pulse is missed because of the propagation delay times of the gates driving one of the gate input pins. Therefore the narrow pulse is gone before the other signal arrives. The result is a lost pulse. This is illustrated by the waveshapes of Fig. 6-4. The upper waveshape is delayed by the gates driving the input pin to a point beyond the narrow pulse. The result is that the narrow pulse is missed.

Another race condition is illustrated in Fig. 6-5. This race condition occurs when one signal changes from one level to another while the other signal changes in the opposite direction. If one pulse is delayed, a narrow pulse may be generated by the overlap, as shown. This narrow pulse may or may not cause a problem, depending on the circuitry that follows. Circuit operation must be analyzed with these conditions in mind.

Table 6-2. Power Requirements Chart for Fig. 6-3

Function	Two Input OR	Two Input NAND	Flip-Flop	
Gates Used	1 2 3 7	4 5	6	Total
Current	22 mA	19 mA	15 mA	46 mA

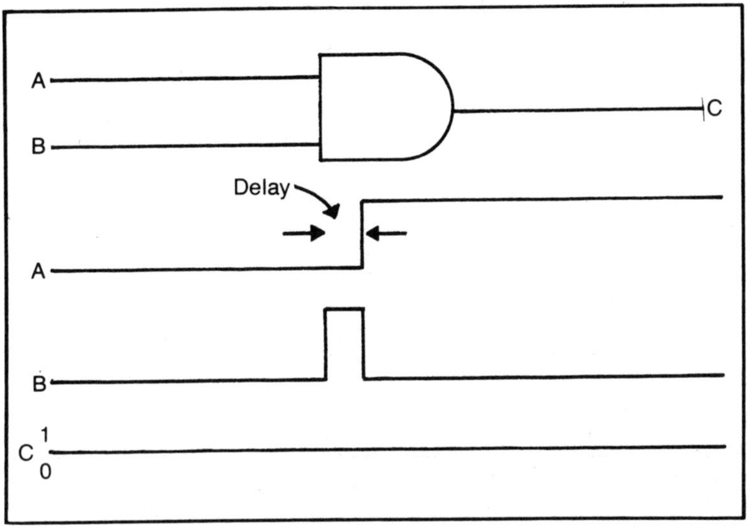

Fig. 6-4. Illustrating the possibility of missing a pulse because of race condition and circuit delays.

Output Loading

Sometimes it is required to drive a load beyond the capability of the standard TTL gate. Depending on the voltage and current required, an open collector driver may meet the requirements. Figure 6-6 shows driving a 30 milliamp, 30 volt load using a 7406 open collector

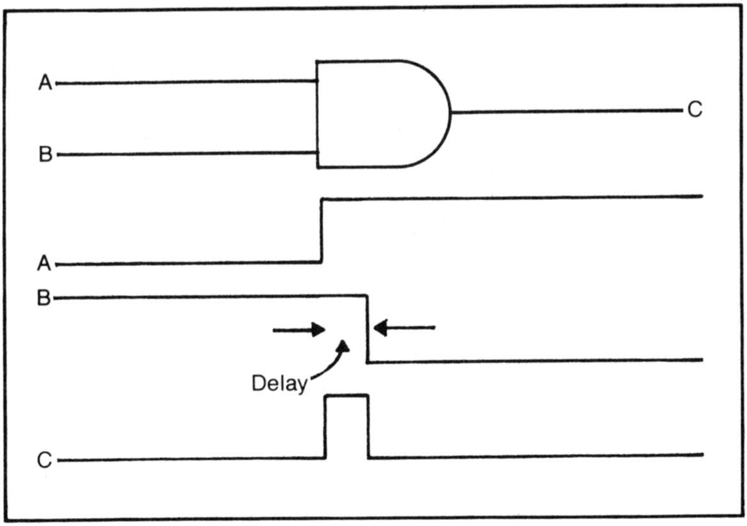

Fig. 6-5. Signal race condition which can generate an extra pulse.

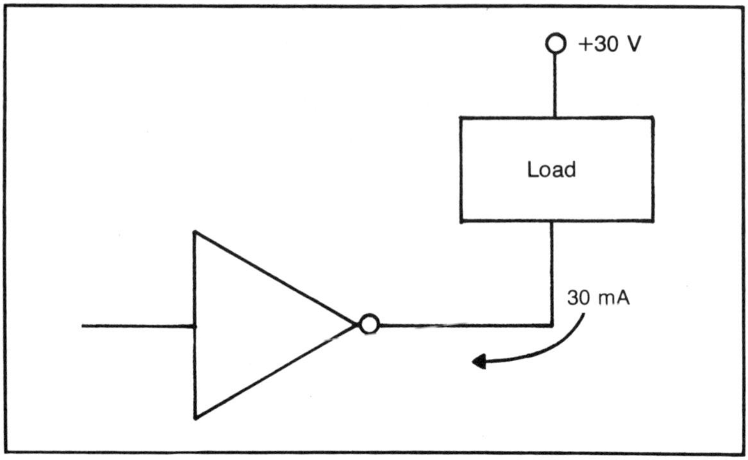

Fig. 6-6. Driving high voltage and high current loads.

driver. This is within the specifications for the driver. Similar noninverting devices are available.

To drive loads of larger current and/or voltage requirements, a transistor or a silicon controlled rectifier may be used. Figure 6-7 shows driving a transistor with a TTL gate. Most of the time the transistor can be directly driven by the IC, as shown. These circuits work for switching the ground on the load, or as a current sink. To work as a current supply, where the high voltage to the load is

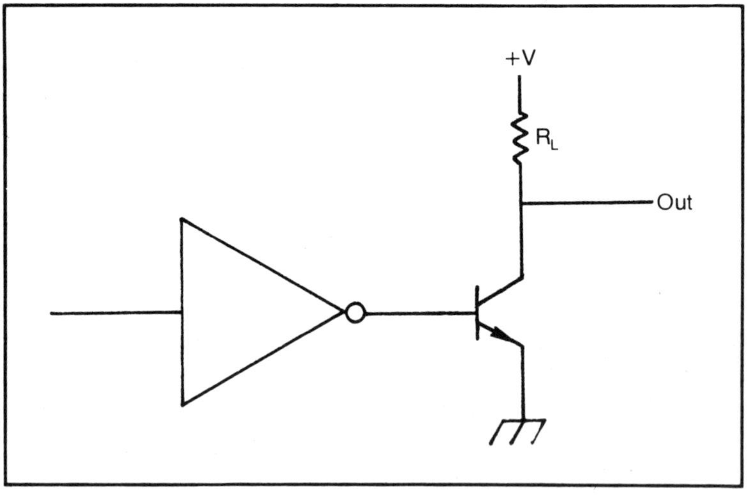

Fig. 6-7. Driving a transistor with a gate to provide increased current and voltage capabilities.

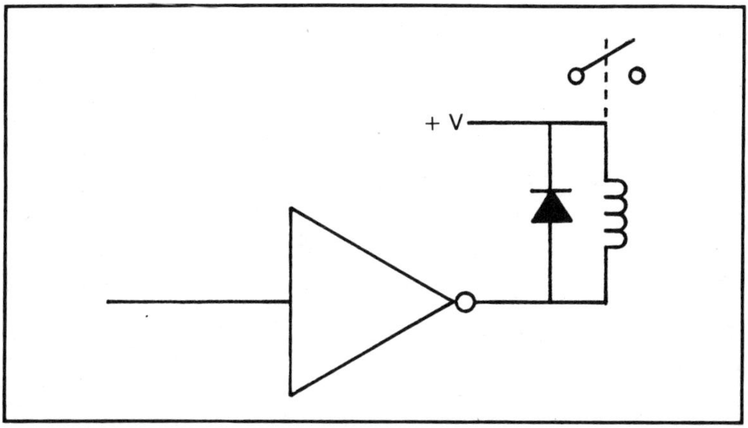

Fig. 6-8. Driving a relay with an open collector inverter.

switched, requires an output which will supply current. Some of the 7400 series will supply some current, but if the requirement is more than these are capable of, a totem pole type transistor circuit may be used.

Another option is to use an analog switch integrated circuit. This is covered in the next chapter.

As a last resort, a relay can be used. Figure 6-8 shows driving a 12 volt relay with a 7406 driver. Relays, because they are high inductive loads, generate a high voltage spike when they are de-energized. This voltage spike can be many times the value of the

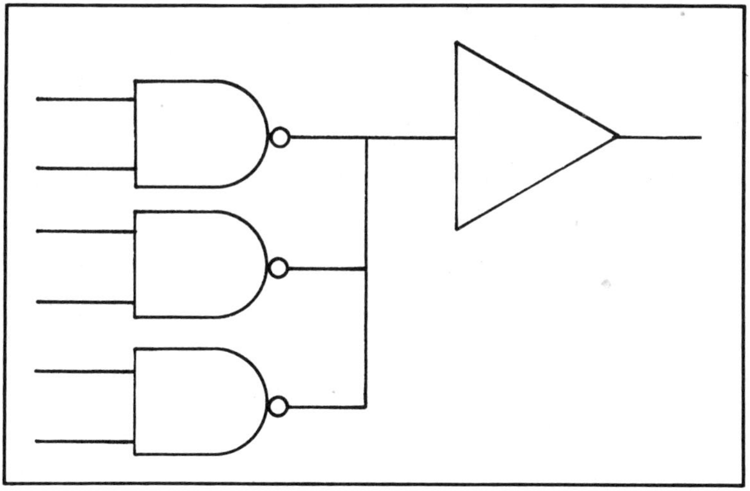
Fig. 6-9. Connecting gates in parallel.

applied voltage, and can severly damage the output circuitry. For this reason some method of spike suppression must be used. The simplest and easiest to use is the diode, connected as shown in Fig. 6-8. This shorts out the voltage generated by the collapsing electric field generated by the relay coil. This method does slow down the relay operation slightly, but it is sufficient for most applications.

Parallel Outputs

The outputs of most TTL gates can be connected in parallel, as

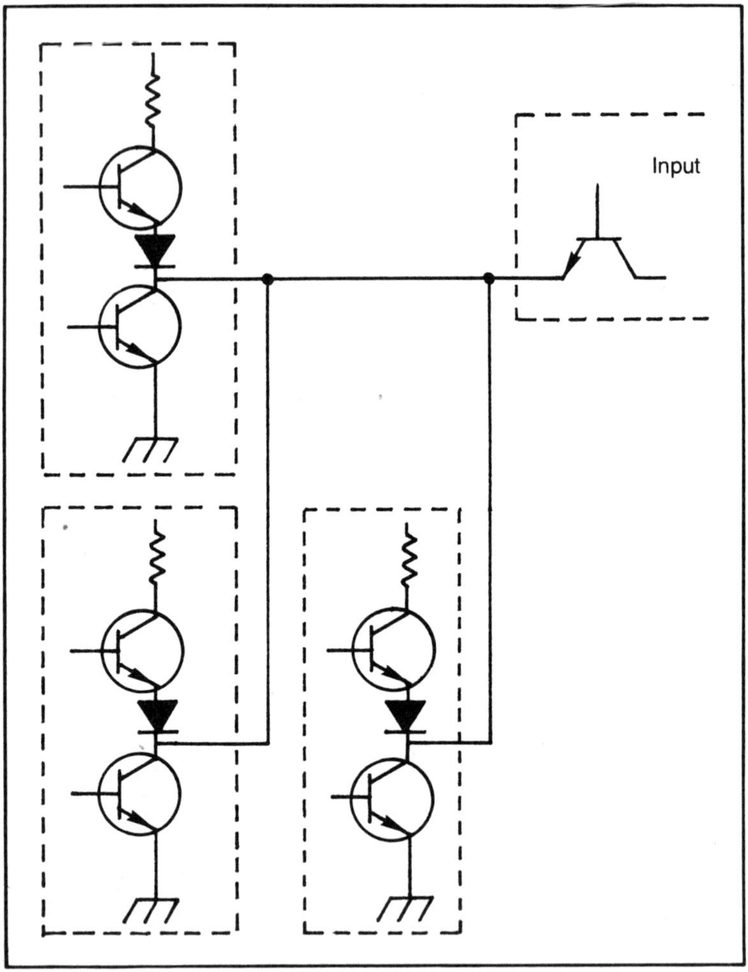

Fig. 6-10. Output circuits of some gates, connected in parallel with the input circuit of the driven circuit, used to illustrate current flow.

shown in Fig. 6-9. This is basically an AND gate because the outputs of all the gates connected to the common point must be high for the point to be high. If any one gate is low, current from the high gate outputs will be pulled low. This condition is illustrated by the circuit in Fig. 6-10. It shows the output stages of the gates connected to a common point. If one of the lower transistors is on, current will flow through that transistor, effectively grounding the common point and the emitter of the input of the gate driven by the common point. This concept can be used in lieu of an AND gate if required. It is advisable to use actual AND gates where possible.

Gates Used as Inverters

Many times it is desired to use an unused gate in an existing integrated circuit as an inverter instead of adding another IC just for one inverter. This can be done by connecting the inputs together. If the drive gate will not allow this, the other inputs can be connected so thay they do not effect the operation. For NAND gates, connect them through a 1 kilohm resistor to+5 volts. For NOR gates, ground them.

Latch Circuit

A latch circuit is commonly used in digital circuits. It remembers the last input on the set terminal until the circuit is reset. A latch circuit is shown in Fig. 6-11. It consists of two 2-input NAND gates connected back to back. More inputs can be used if required by the circuit. If output C is the active low output, then output D is the active high output, input A is the active low reset input, and input B is the active low set input. As long as both inputs are high, the state of the latch outputs will not change. To analyze this circuit, assume that output C is high and output D is low. Also let both inputs be in the nonactive, or high state. Since D is low, and A is high, the output of the upper NAND gate is high. This high at C, and the high at B result in the output of the lower NAND gate being low. Thus status quo is maintained.

If input B is changed to low, then the output of the lower NAND gate will be high, making output D high. This, and the high at A result in both inputs to the upper NAND gate being high, giving a low output at C. This low will latch the lower output high. Thus the outputs have changed state.

If input B is changed back to high, the circuit will maintain status quo until input A is made low. This low makes the output of the upper NAND gate and output C high. The high at C and the high

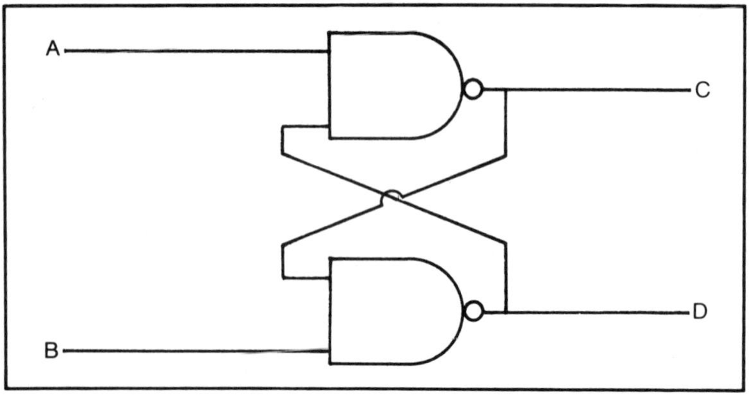

Fig. 6-11. The latch circuit made from two NAND gates.

at input B, will result in a low output at D. This low feeds back to the upper NAND gate, latching the output high. Thus the outputs have changed states again, and returned to the original condition. The waveshapes of Fig. 6-12 illustrate the operation of this latch. The truth table given in Table 6-3 is for this circuit. This truth table is a sequential truth table, meaning that one step is based on the previous step being completed. In this truth table, it is especially true for lines 2 and 4.

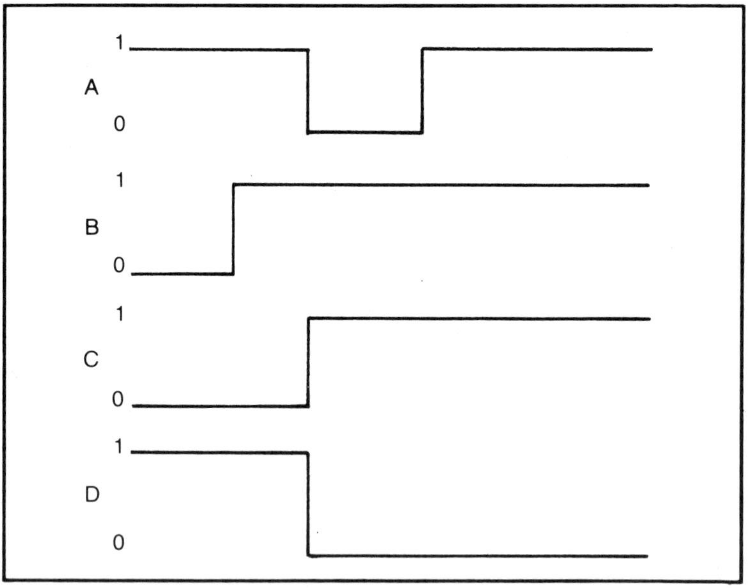

Fig. 6-12. Waveshapes for the latch circuit.

A	B	C	D
1	0	0	1
1	1	0	1
0	1	1	0
1	1	1	0

Table 6-3. Truth Table for the Latch Circuit

In short, the latch will assume the set condition (the set output is active high) when the set input is active. It assumes the reset condition when the reset input is active. If both inputs are in the nonactive condition, the latch will maintain the previous outputs.

Active Level

When analyzing or designing a digital circuit the *active level* must be determined and remembered. This active level changes from point to point along the circuit. A simple inverter changes the active level from low to high, or visa versa. Look at the truth tables for the NAND or NOR gates given in Table 6-4. The inputs are active high while the outputs are active low. The active state is the state that carries the information. It signifies the operation has been accomplished, the pulse present, or the condition met. A low from a NAND gate indicates that the input logic conditions have been met, while a high from an AND gate indicates the same thing. This is a common problem. The active state is in the wrong level at a point in the circuit.

Returning again to Table 6-4, look at the operation of the NAND and NOR gates for active low inputs. For the NAND gate, when any input is low, the output is high. This is the requirements

Table 6-4. NAND and NOR Truth Tables

NAND		
A	B	OUT
0	0	1
0	1	1
1	0	1
1	1	0

NOR		
A	B	OUT
0	0	1
0	1	0
1	0	0
1	1	0

for a NOR gate, so the NAND gate becomes an active low input NOR gate. The same applies to the AND gate. It becomes an active low input OR gate. For the NOR gate, the output is high only when both inputs are low, so it becomes an active low input NAND gate. An OR gate becomes an active low AND gate. So, if the signals are active low, it is advantageous to consider the gates as active low input gates. This may eliminate the need to invert signals to obtain the proper active level.

Clock

Some logic circuits operate with a clock to sequence signals through the system. This clock pulse steps the information through the circuit one, or more, steps at a time. Depending on the circuit complexity and the gate time delays, several gates may be operated on one clock pulse. The clock pulse is used to enable the gate and the information applied to other inputs. Typically, this is accomplished using NAND and AND gates with the clock connected to one of the inputs.

Truth Tables

Now, let us look at a more complex logic circuit and analyze it with a truth table. The circuit is shown in Fig. 6-13. The end result is to set the latch if A and B, or C and D, along with E or F, are high. Looking at the circuit, it is hard to visualize if the circuit is operating correctly for all combinations of inputs.

Therefore, construct a truth table for this circuit. Start with all

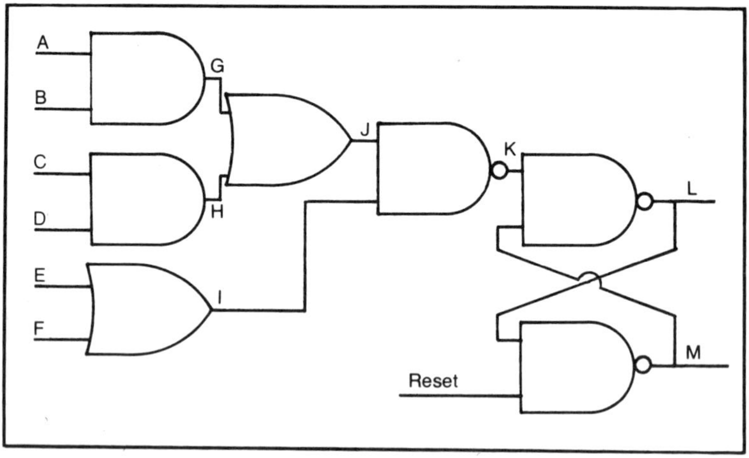

Fig. 6-13. Logic circuit for the truth table of Table 6-5.

Table 6-5. Partial Truth Table for Fig. 6-13

A	B	C	D	E	F	G	H	I	J	K	Reset	L	M
0	0	0	0	0	0	0	0	0	0	1	0	0	1
1	0	0	0	0	0	0	0	0	0	1	1	0	1
0	1	0	0	0	0	0	0	0	0	1	1	0	1
1	1	0	0	0	0	1	0	0	1	1	1	0	1
0	0	1	0	0	0	0	0	0	0	1	1	0	1
0	0	0	1	0	0	0	0	0	0	1	1	0	1
0	0	1	1	0	0	0	1	0	1	1	1	0	1
1	0	1	1	0	0	0	1	0	1	1	1	0	1
0	1	1	1	0	0	0	1	0	1	1	1	0	1
1	1	1	1	0	0	1	1	0	1	1	1	0	1
0	0	0	0	1	0	0	0	1	0	1	1	0	1
1	0	0	0	1	0	0	0	1	0	1	1	0	1
0	1	0	0	1	0	0	0	1	0	1	1	0	1
1	1	0	0	1	0	1	0	1	1	0	1	1	0
0	0	0	0	0	0	0	0	0	0	1	0	0	1
0	0	1	0	1	0	0	0	1	0	1	1	0	1
0	0	1	1	1	0	0	1	1	1	0	1	1	0
0	0	0	0	0	0	0	0	0	0	1	0	0	1
1	1	1	1	1	0	1	1	1	1	0	1	1	0
0	0	0	0	0	0	0	0	0	0	1	0	0	1
1	1	0	0	0	1	1	0	1	1	0	1	1	0
0	0	0	0	0	0	0	0	0	0	1	0	0	1
1	1	1	1	1	1	1	1	1	1	0	1	1	0

inputs, and the reset low. Signals G, H, and I will be low, so J will be low and K will be high. K high and reset low will set output L low and output M high. To change the states from this, the reset must be high and K must be low. For K to be low requires that I and J both be high. For I to be high either E or F or both must be high. For J to be high, either G or H, or both must be high. For G to be high, A and B must both be high. For H to be high, C and D must both be high. This is basically how to set the circuit, but only implies what happens for the other input conditions.

To construct the truth table as shown in Table 6-4, change the inputs and determine the condtions for each change. Continue to do this until all the different conditions for the inputs and the reset are met. Notice that the truth table is again a sequential table. Whenever the circuit is set, the next step is to apply a reset input, to reset the circuit for the next input.

There are several steps missing from the truth table, mainly in the lower portion of the table. You can determine which steps are missing and make up a truth table to include them. This would be good practice.

FLIP-FLOPS

Another common type of digital IC is the flip-flop. Flip-flops are similar to latches in that they maintain a condition until they are triggered to the other condition. There are several types of flip-flops, such as the RS type, the JK type, and the D-type edge-triggered. It is the method of triggering which separates the types. The general logic symbols for the three types of flip-flops are shown in Fig. 6-4.

The RS flip-flop is a simple set/reset type, whose operation is similar to that of a latch. The set and reset inputs may be active high or active low, depending on the type number. When the set input is active, the circuit is forced to the set condition with output Q high and output $\overline{Q}$ low. This condition is maintained until the reset input goes active. Then the circuit is forced into the reset condition, with the Q output low and the $\overline{Q}$ output high. It will remain in this condition until the set input goes active. In this manner, the last input is remembered. If both the set and reset inputs are active at the same time, the circuit will not change states.

The JK flip-flop has two data inputs, J and K. The J and K inputs are loaded into the flip-flop when the clock is active. The J and K inputs control what data is loaded into the flip-flop while the clock controls when the data is loaded. When the J and K inputs are both active, the output of the flip-flop will not change. When only one of them is active, the flip-flop will load that condition when the clock goes active. Some JK flip-flops have a master set and/or master reset inputs. These inputs will force the flip-flop into the set or

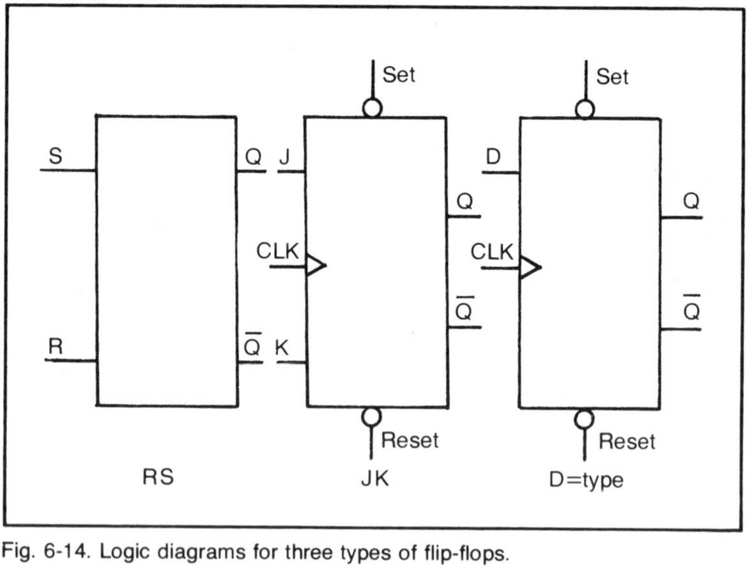

Fig. 6-14. Logic diagrams for three types of flip-flops.

reset condition, no matter what the condition of the JK inputs and the clock input.

The D-type edge-triggered flip-flop also uses the clock input to trigger the flip-flop. For example, the 7474 dual D-type flip-flop will load the data at the D input into the circuit at the low-to-high transition of the clock. This data appears at the Q output, while the $\overline{Q}$ output is inverted. The flip-flop will stay in this condition until the next low-to-high transition of the clock, or until the set or reset inputs are enabled. Typically, the set and reset inputs are active low so that a low on the set input will drive the flip-flop to the set condition (Q high and $\overline{Q}$ low). Likewise, a low on the reset input will drive the flip-flop to the reset conditon (Q low and $\overline{Q}$ high). The set and reset inputs have priority over the data and clock inputs. Even if a low-to-high transition of the clock input occurs while either the set or reset inputs are low, the circuit will not recognize the data input.

Once the flip-flop has been set or reset and the set and reset inputs are not active, the circuit will remain in that condition until some input causes a change. The only inputs that can cause a change are a set input low and a reset input low, or a low-to-high transition on the clock input. So the flip-flop will remain in the last state until forced into the other state. The set and reset inputs are used to force the flip-flop to a known condition, usually at the start of a logic operation. Typically, logic circuits are reset at power up by the system reset. This gives the circuit a known starting point.

The D-type flip-flop can be used as a set/reset flip-flop by using only the set and reset inputs. But the standard RS flip-flop is better suited for this application.

Uses

There are several uses for the flip-flop, with the most prevelant being as a memory element. It can remember the passage of a condition or pulse after the condition or pulse has passed. It can also remember information loaded into it in binary form, such as one binary bit of a binary word. To remember a 4-bit binary would require four flip-flops.

In the very early days of computers, flip-flops were used as memory elements. Especially in such areas as registers, adders, central processing units, and temporary memory. Although they were complex and expensive as compared to other memory elements, they found widespread use. They are still used extensively as a memory to remember the state of a logic signal after it has passed, such as registers and shift circuits.

A flip-flop can remember information from a current logic sequence to be used in a later logic sequence, thus simplifying the timing and eliminating problems due to differences in propagation times.

Another use for flip-flops is in countdown circuits. These countdown circuits count down the input clock pulse and generate a sub-multiple of the pulse rate. Since the flip-flop will only change states on the low-to-high transition of the clock the output will change states once for every full input pulse if the D input is the inverse of the Q output. In other words, two input pulses can give one output pulse. The $\overline{Q}$ output is the inverse of the Q output, and the Q output is the same as the D input at the low-to-high transition of the clock pulse. Connecting the $\overline{Q}$ output to the D input will enable the flip-flop to change states once everytime the input (clock input) changes state twice. The low-to-high transition of the clock input will cause the output to change state, but the high-to-low change of state of the input will not cause the output to change state because it is not one of the trigger conditions. The result is that the input pulse rate is divided by two.

Figure 6-15 shows the waveshapes for the 7474 D-type flip-flop, when used in the normal mode. This waveshape gives the basic operation of the flip-flop.

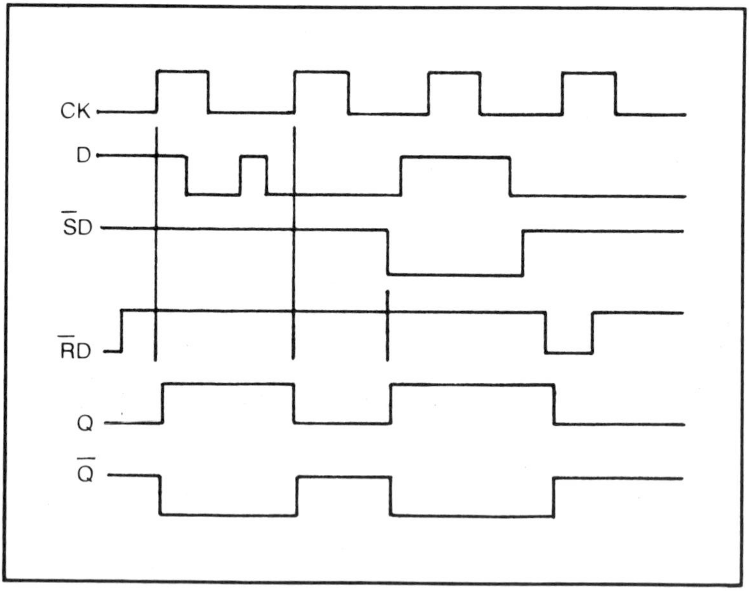

Fig. 6-15. Waveshapes for a typical D-type flip-flop.

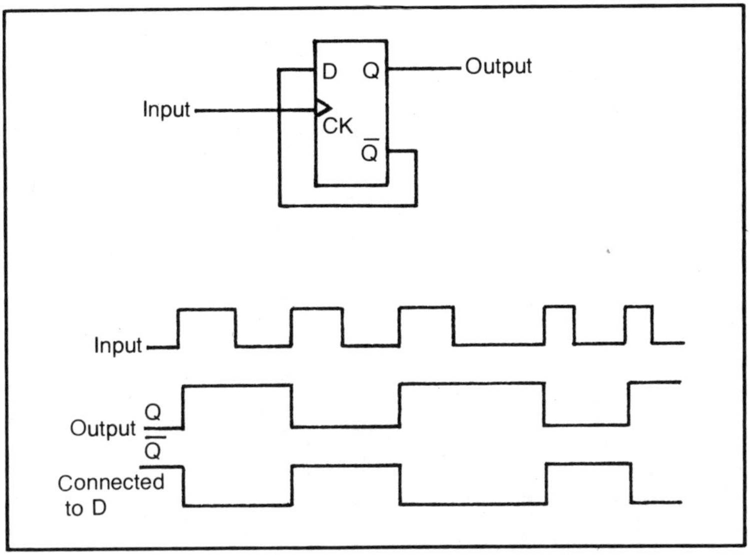

Fig. 6-16. The 7474 flip-flop connected as a divide-by-two countdown circuit and the associated waveshapes.

Flip-Flop

A 7474 flip-flop connected as a countdown circuit (divide-by-two) is shown in Fig. 6-16 with the associated waveshapes shown underneath. It is assumed that the set and reset inputs are high. Study the waveshape and the circuit until the correlation between the two is understood, and the operation of the circuit can be followed using the waveshapes.

Several flip-flops, connected in the divide-by-two configuration can be connected in series to provide a larger countdown. Each stage will have a divide-by-two result, so the end result will be two raised to some power (the number of stages). Figure 6-17 shows such a connection with four countdown circuits connected in series. The associated waveshapes are shown below the circuit. This circuit is a divide-by-16 circuit, with 16 input pulses required to count the circuit back to the starting point, with all the outputs zero. When 15 pulses have been received, the outputs of all the stages are high, or ones. The next pulse (pulse 16) will cause the count to start over with all the outputs zero.

From this circuit and the associated waveshapes, any one pulse can be derived with an AND gate. If a 4-input AND gate is used, as shown on the left side of Fig. 6-18, the length of the output pulse will be the length of the shortest of the output pulses from the stages. So

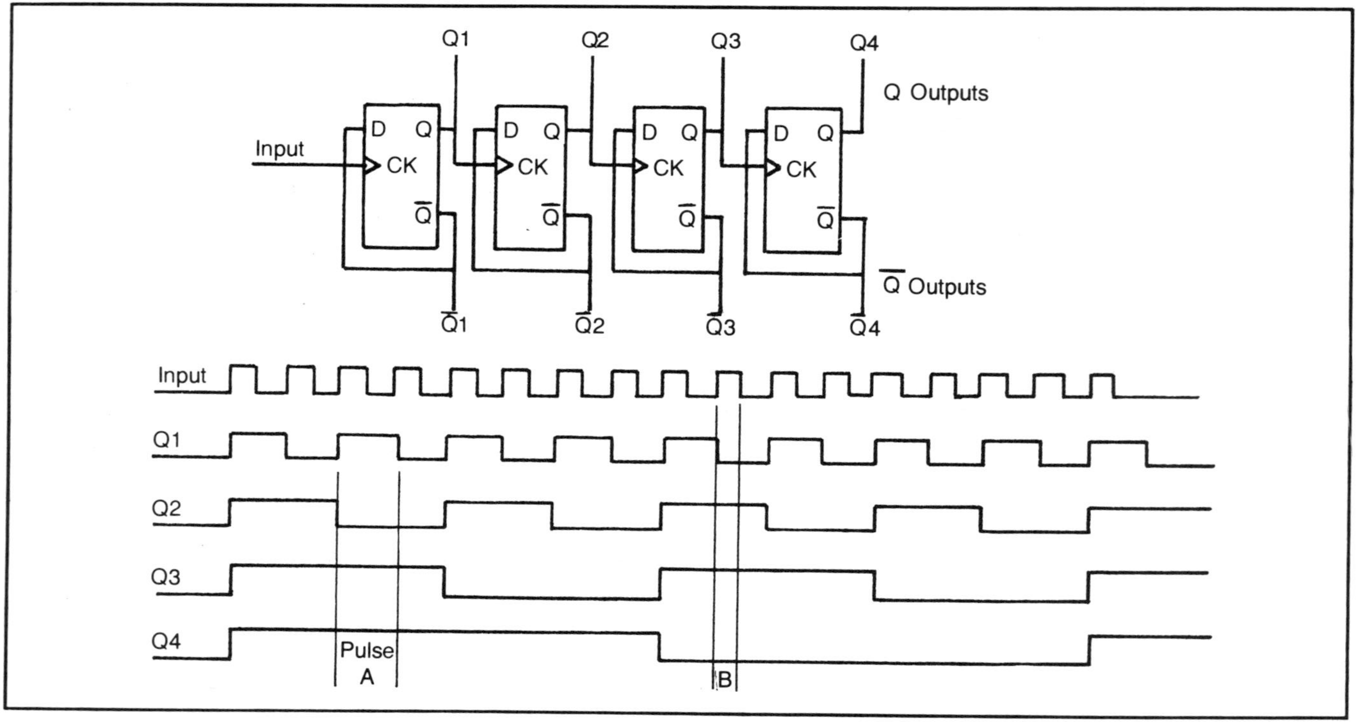

Fig. 6-17. Four 7474 flip-flops connected as a four stage countdown circuit and the associated waveshapes.

the length will be the length of the pulse from stage 1, which is the shortest pulse as can be seen in Fig. 6-17. Using a 5-input gate, and connecting the input to one of the inputs, the output pulse width will be that of the input to the first stage. This is shown by the circuit on the right side of Fig. 6-18. The inputs to the left circuit of Fig. 6-18 are connected to develop the pulse at time period A shown on the waveshapes of Fig. 6-17. The right circuit is to generate a pulse during time period B of Fig. 6-17. The inputs to the AND gates connect to the outputs of the divide-by-16 circuit as notated.

From Fig. 6-17, the time period for pulse A requires that Q1, Q3, and Q4 be high and Q2 be low, so the AND gate is connected to the Q1, $\overline{Q2}$, Q3, and Q4 outputs. Likewise the pulse at time period B requires that Q1 and Q4 be low, while Q2 and Q3 are high, all during the high portion of the input. The logic equations for these two circuits are as follows:

$$A = Q1 \cdot \overline{Q2} \cdot Q3 \cdot Q4$$
$$B = \text{input} \cdot \overline{Q1} \cdot Q2 \cdot Q3 \cdot \overline{Q4}$$

Consider the next half cycle of the input after pulse B, where the input is low. If a pulse is required during this time period, an inverter may be used on the input because there is no $\overline{\text{input}}$ signal available. Another possibility is to use a NAND or a NOR gate. As an exercise in logic circuits, you should determine which is required, and the proper connections for this time period.

To extend the countdown circuit to additional stages, simply connect more divide-by-2 circuits to the Q output of the last stage. Realize that for each added stage, the total divide-by will be doubled. Adding one stage to Fig. 6-17 will make a divide-by-32 circuit.

To make a circuit which counts by some other number, such as a divide-by-8 or divide-by-14, connect an AND gate to reset the

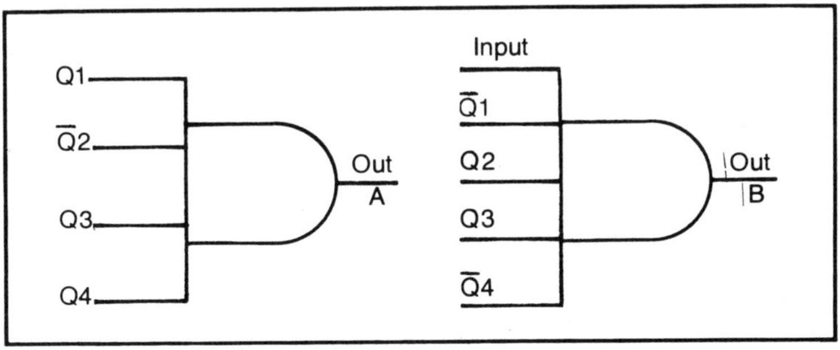

Fig. 6-18. Using AND gates to generate individual pulses from Fig. 6-17.

flip-flops when the count is reached. By connecting the output of the above AND gates to the reset input of all the flip-flops, the circuit will reset when the pulse is reached and the count will start over at all zeros.

Another use for flip-flops is in shift registers. A shift register shifts in data serially and reads it out in parallel. Serial data is where one data bit follows another in one waveshape, as shown in Fig. 6-19. The upper waveshape defines the bit times, or times when the data is valid and can be sampled. The second waveshape is of the return to zero format. Data returns to zero between each data bit, so for one's, there is a pulse, and for zero's, there is no pulse. The third, or lower waveshape shows the nonreturn to zero format. In this format, the data waveshape does not return to zero between data bits. The only time the data waveshape is zero is when there is a zero for the data bit. This waveshape has pulses only when the data goes from a zero to a one (positive going edge of the pulse) and then back to a zero (negative going edge.)

Serial data requires only one line on which data is carried. This reduces the number of wires and connections, and is used quite a bit for interconnection between equipment where logic signals and data words are passed back and forth. But serial data transmission takes time for all the bits to be operated on by the circuit, because they must be processed one bit at a time. It also requires timing and gating of the sequential pulses.

Parallel data flow is where all the bits which comprise the data word are carried on parallel lines, during one bit time. Figure 6-20 shows a parallel 4-bit data waveshapes. This figure shows four 4-bit data words, with each waveshape representing one of the bits of the data word. This type of data handling is faster and has less timing requirements, but it takes more components and more wires. Present day computers are parallel data flow because of the speed. This

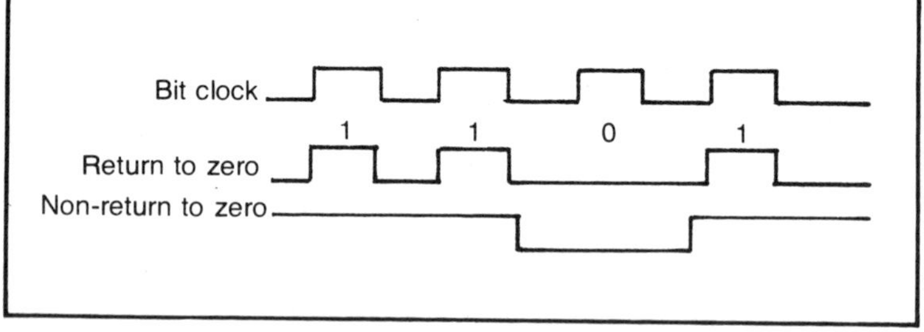

Fig. 6-19. Serial data bits of both the return-to-zero and the nonreturn-to-zero types.

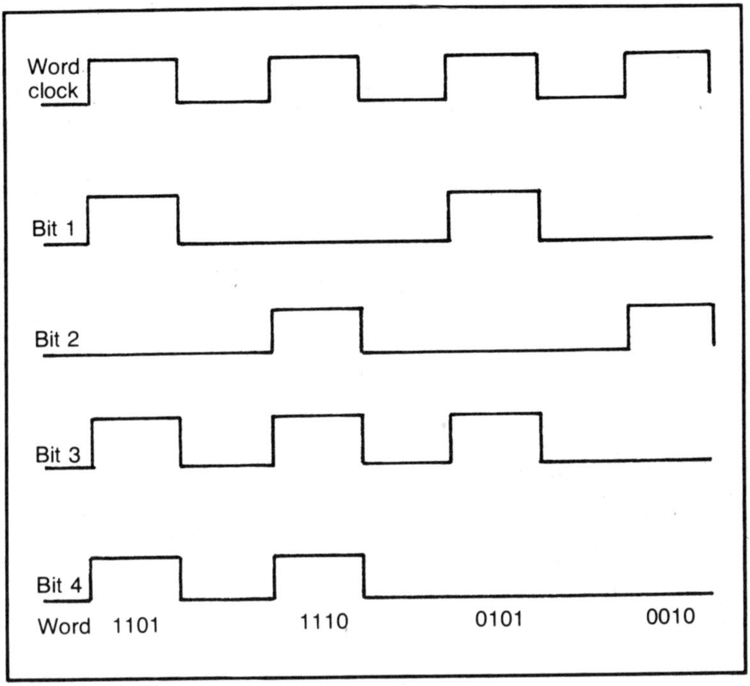

Fig. 6-20. Parallel data flow showing four 4-bit data words and the word clock.

is a subject which is covered in the introduction to computers, and is beyond the scope of this book.

One data bit is one pulse or level, sampled at the bit time. A 4-bit data word is 4 data bits. The size of the data word is dependent on the logic circuits involved, and can be any number of bits. But the number of bits comprising a data word is fixed for a given circuit.

The function of the shift register is to convert from the serial data to parallel data. A typical application is at the interface between a logic system and the control panel. Figure 6-21 shows a 4-bit shift register for converting a 4-bit serial data word to a 4-bit parallel data word. All the flip-flops are initially reset, so the outputs are all initially low. The first data bit appears at the input of the first flip-flop, and at the transition of the clock pulse, this data is loaded into the first flip-flop. It appears on the Q output. At the second transition of the clock pulse, the first data bit is moved to the second flip-flop, and the second data bit is loaded into the first flip-flop. After the fourth data bit has been loaded by the fourth clock transition, the data word is loaded into the shift register, and the outputs represent the parallel data. The parallel data must be read by the

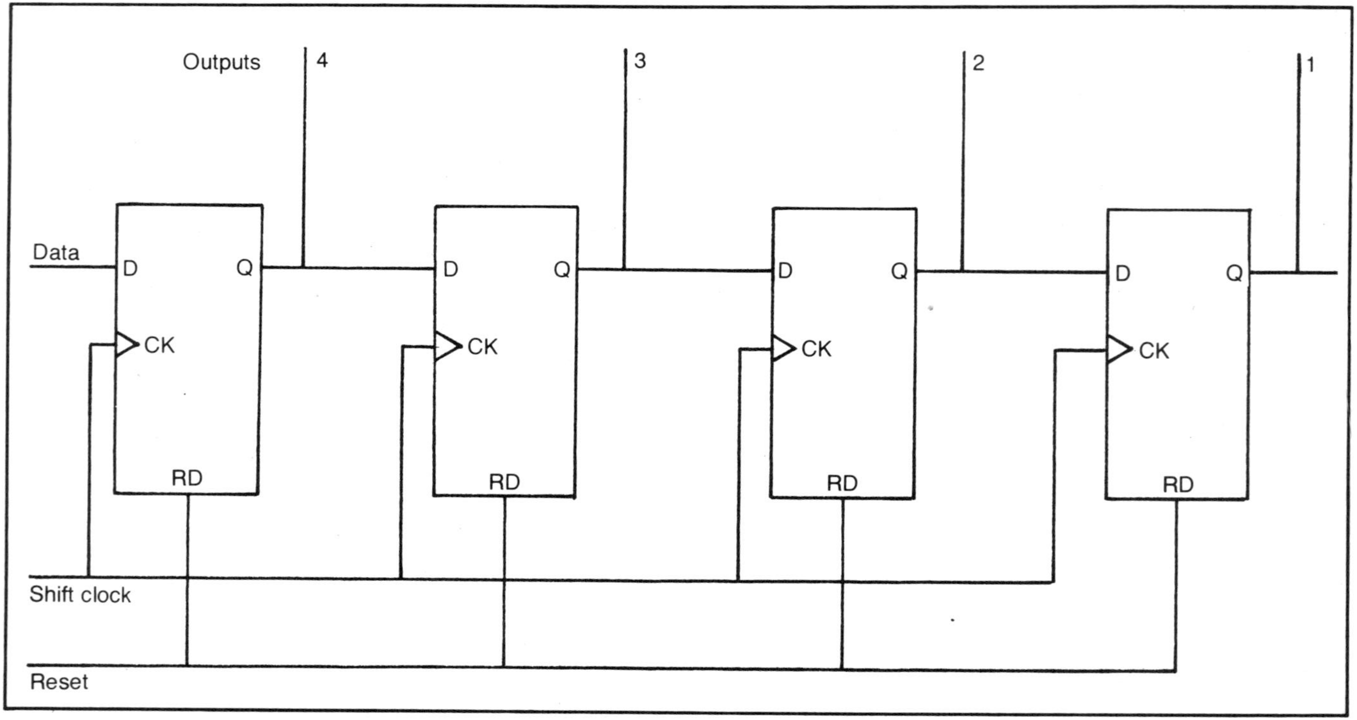

Fig. 6-21. A 4-bit shift register using 7474 flip-flops.

logic circuit before the next clock transition, because this will shift in another data bit.

COUNTER MODULE

Next, let us look at the 7447 seven-segment decoder/driver, the 7490 BCD counter, and the seven-segment display together. These three devices are used together quite often. The 7447 is used to decode BCD data and generate the respective segment drive currents for a seven-segment LED display. The 7490 is a decade counter which generates parallel BCD outputs for a serial pulse train input. Although this IC has many other uses, one of the primary uses is with the decoder and display to indicate a decimal count.

Seven-Segment Displays

Seven-segment LED displays are numeric displays made up of seven LED segments in the shape of bars. Figure 6-22 shows a LED seven-segment display, with the segments identified with the standard lettering. Also shown are the segments required to make up the numbers, and what the display looks like. It is a series of short, straight lines which are chosen to display the number. Each segment is connected to a pin on the package case, so activating that pin will illuminate that segment.

There are two basic types of LED displays, the common anode and the common cathode. This refers to which end of the individual LED segments is common. The most common for use with TTL logic circuits is the common anode. This is because the positive connection for the LEDs is connected to the VCC and the negative ends of the LEDs are driven by the external logic circuits. This requires a ground at the pins to turn on the segment, and TTL circuits drive the ground, or low output.

As with all LEDs, LED displays are current devices. This means that it is the current which operates the LED, and the voltage drop across the LED is low. This also means that the on condition of the LED is a low impedance state. The current must be limited to a value below the maximum current by external resistors. For the average, standard size display, this maximum current is about 16 mA. Excessive current will cause the LED to become bright and then burn out. Typically, the current is limited to 5 mA to 10 mA, depending on the intensity required. The method of limiting the current is to connect a resistor between the segment pin of the display and the corresponding output pin of the driver.

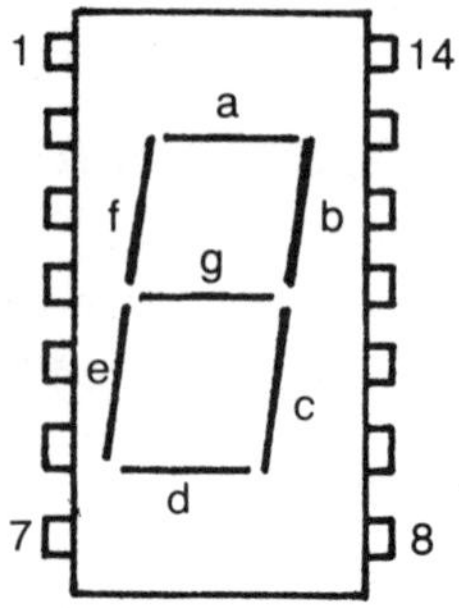

Number	Display	Segments
0	0	a, b, c, d, e, f,
1	1	b, c
2	2	a, b, g, e, d
3	3	a, b, g, c, d
4	4	f, g, b , c
5	5	a, f, g, c, d
6	6	f, e, g, c, d
7	7	a, b, c
8	8	a, b, c, d, e, f, g
9	9	g, f, a, b, c

Fig. 6-22. An LED seven-segment display.

For a common anode type display, the active state of the inputs is low, so the resistor limits the current flow through the LED to the driver and on to ground. The current limiting resistor cannot be connected in the common lead, or between the common anode pin and the +5 volts. A resistor connected there will cause the intensity to decrease as more segments are activated. If the segment current is limited to 16 mA with one segment on, then it will be 8 mA with two segments active, 4 mA with four segments active, and

2 mA with all segments active. This will give quite a difference in intensity.

The series current limiting resistors must be connected between the segment pin and the driver output pin. This will keep the segment current constant. Consult the specifications for the type number used to determine the safe operating region. Some display types have internal series current limiting resistors, and external resistors are not required.

Most seven-segment displays also contain a decimal point, designated as DP. This decimal point may be before or after the numeric display, depending on the type number. Again, consult the specifications to determine the location of the decimal point.

Several different seven-segment displays are available. They are available in different sizes, colors, and pin configurations. The most prevalent color is red, although green and amber are available. Most of the displays have the standard pin spacing on the bottom to match the 14-pin and 16-pin DIP packages. This makes it very convenient to mount the display. Simply use the standard 14-pin or 16-pin DIP socket. Always check the pin definition for the display you are using to verify the pin numbers.

Figure 6-23 shows the schematic for a counter module using

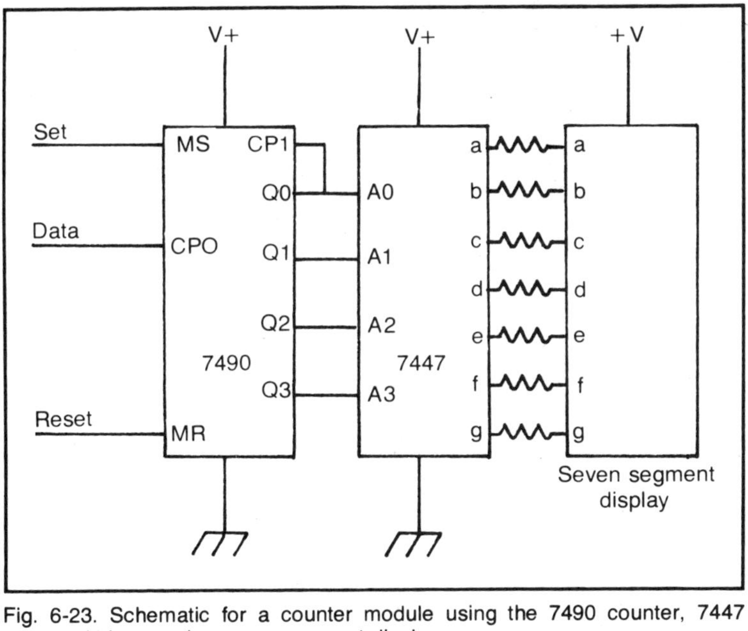

Fig. 6-23. Schematic for a counter module using the 7490 counter, 7447 decoder/driver, and a seven-segment display.

the 7490, 7447, and a seven-segment display. This forms a simple, one digit decimal counter which can be incorporated directly into logic circuitry.

The series limiting resistors are connected between the segment inputs of the display and the segment outputs of the decoder/driver. The resistors that I used for this circuit are 470 ohms, and they limit the LED current to about 9 mA, which is sufficient brightness. Remember, for the 7490 counter to count pulses, at least one set and one reset input must be low.

7490 Counter

Note that the 7490 counter does not have a carry, or overflow output. This type of output generates a pulse for the next digit when the counter goes back to zero after a maximum count. If more than one digit is required, the carry must be generated externally. One easy method of overcoming this shortcoming is to use a 74192 decade counter. It has a carry output built into its circuitry. A four input NOR gate connected to the four outputs of the 7490 will generate a high output only when all of its inputs are low. This condition will occur when the counter goes from its maximum count of 9 to zero. The output of the NOR can be connected to the clock in for the next counter. Therefore, when the first counter has completed a full count it resets to zero and the NOR gate increments the next counter. Figure 6-24 shows the waveshapes for the counter and the NOR gate.

An addition to a counter circuit that is usually required is a reset switch. The switch manually resets the counter. The reset switch can be a pushbutton switch that is normally closed. Depressing the switch will open the contacts. A reset switch and the associated circuitry is shown in Fig. 6-25. When the switch is closed the reset input is grounded, allowing the counter to count. When the switch is depressed and the contacts are opened, the high bias resistor places a high level on the reset input, resetting the counter. Typically, logic control of the reset input would also be provided. This allows the system reset to also reset the counter and allows other logic functions to reset the counter.

Another typical application for the 7490 is shown in Fig. 6-26. This circuit is a serial to parallel converter which converts a serial input pulse count to a 4-bit BCD parallel count. This type of circuit is used to count the pulses during the time the run signal is active. It is also used to generate pulses during a given time period or at a

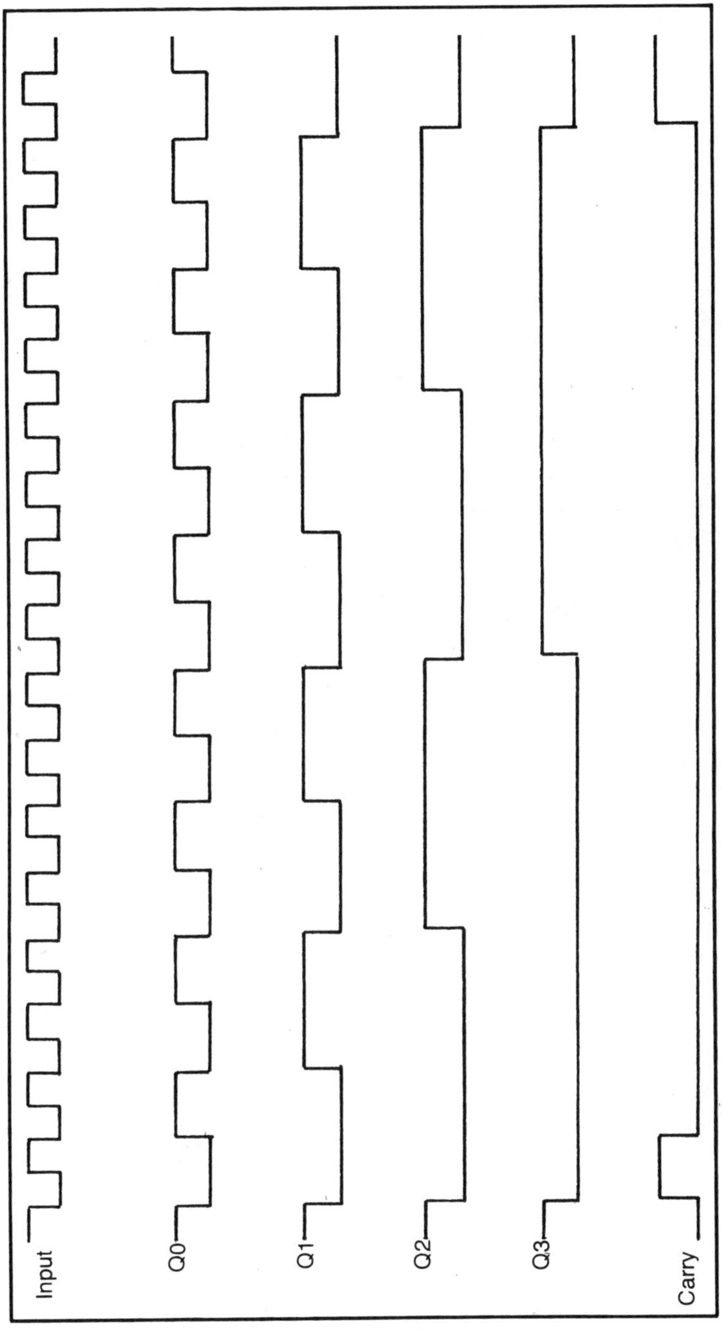

Fig. 6-24. The waveshapes for the counter module and carry output.

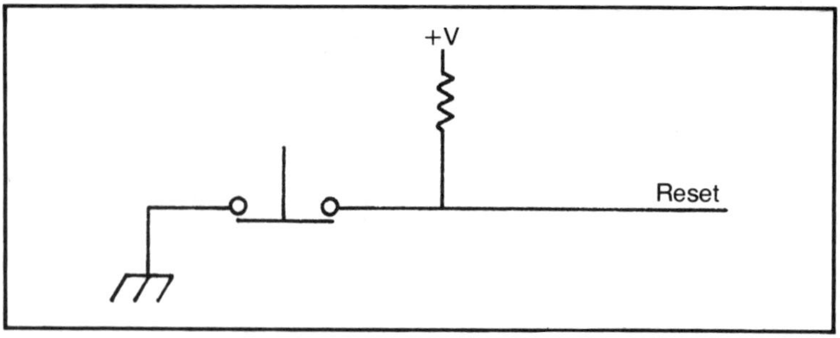

Fig. 6-25. The reset switch for resetting the counter modules.

given pulse number. If the input is a clock pulse, then this circuit counts clock time.

The AND gate connected to the input of the 7490 counter is controlling when the circuit is counting and when it is inhibited. If the run signal is low, counting is inhibited. When it is high, the input pulses will appear at the input CK0 of the 7490. The reset input allows the 7490 to be reset before starting a count. Remember that one of the set inputs must be grounded, or at a low level, to allow the 7490 to count. Refer to the mode selection truth table for the 7490 decade counter.

This is a different type of serial-to-parallel converter than that discussed previously. That converter converted serial data pulses to parallel data. This converter converts a series of timing pulses to a parallel count. By using an AND gate and some inverters, any one of the 10 time periods can be isolated and a pulse generated during

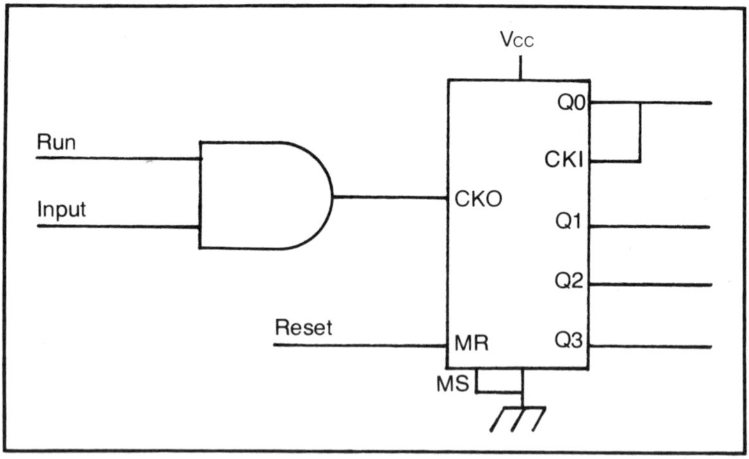

Fig. 6-26. Serial to parallel conversion using the 7490.

that time. For example, the circuits shown in Fig. 6-27 show how to generate a pulse in bit time 0 and in bit time 7. Inverters are used to obtain the not conditions, where required. Consult the BCD-to-binary conversion tabel (Table 5-17) to determine what the state of each output must be to obtain the desired bit time. For time 0, the binary equivalent is 0000, or all outputs in the low state. Follow through the left circuit of Fig. 6-27 to assure yourself it will produce a positive pulse at bit time zero. This method requires the use of inverters on all the AND gate inputs. This being the case, it is possible to combine the functions and use one gate to accomplish the results. Remember from the discussion about NOR gates, that the output of a NOR gate is low whenever any one of the inputs are high. The output is high when all the inputs are low. These are the same as the requirements for this circuit to produce a pulse at zero bit time. So, using a NOR gate in place of the AND gate and inverters will save some gates in the circuit.

For bit time 7, binary equivalent is 0111, requires that the output from counter stage 4 be low, and the remaining counter outputs high. So this requires only one inverter on the Q3 input to the AND gate. Any of the bit times can be generated in this fashion. If the pulse, say at time 7, is generated and the counter allowed to run till full and restart, then the full count will be 16, and the pulse will be at time 7. But if the pulse at time 7 is used to reset the counter, then the count will start over from zero. This shortens the length of the count for the counter.

From the previous examples, you can see that there is usually

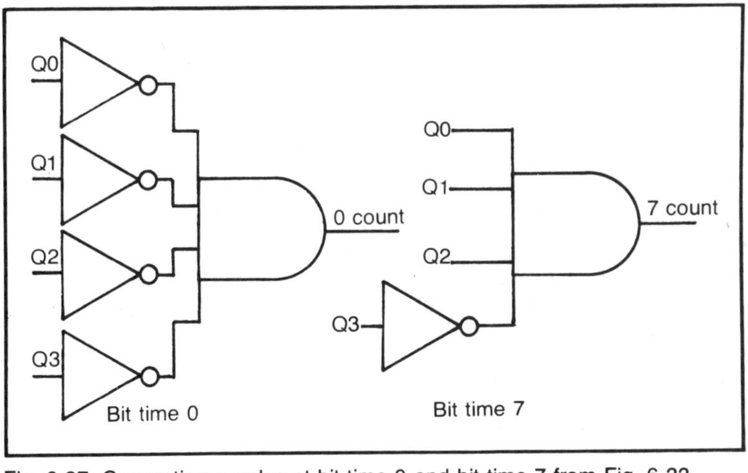

Fig. 6-27. Generating a pulse at bit time 0 and bit time 7 from Fig. 6-33.

more than one method to accomplish the desired logic function. The first approach you chose may not always be the best or the easiest to implement. Always try to look at alternative methods when building logic circuits. There are methods for simplifying logic circuits using logic equations and some types of logic maps. But these are for the more complex logic circuits, and are beyond the scope of this book. For now, it is sufficient to look at the various gates and their functions and see if the function can be accomplished with fewer gates in a simpler circuit. Typically, the gates used are those the builder is familiar with and has on hand. For this reason, it is advisable to become familiar with as many different types of gates as possible.

COMPARATOR

The 7485 4-bit magnitude comparator is used in logic circuits. It compares two 4-bit words to determine their relationship. This IC can determine if the two words are equal and distinguish which of the two is larger. One application is in the set up of an automatic sequencing system. The system sequences several different types of printed circuit boards to test for proper operation. The printed circuit card code number is read into the system from programming switches. Then the stepping switches, which set up the system, step around till the code from the stepping switches matches the programmed code. When the two codes (four bit binary words) are equal the stepping stops. The greater than and less than outputs are used to tell the stepping switch which way to step.

Another use of the 7485 comparator is to give an output when a particular bit pattern appears on the parallel data or output lines of a computer. This application usually uses two 7485 comparators in cascade to compare 8-bit computer words. When the particular word appears, the computer control logic knows that some other function is to be done.

The expansion inputs of the 7485 comparator allow for comparing binary words larger than 4 bits by using more than one IC. The expansion inputs are identified as $I_{A>B}$, $I_{A=B}$, and $I_{A<B}$ and are the least significant bit positions. When used for series expansion, the A>B, A=B, and A<B outputs of the least significant word are connected to the corresponding $I_{A>B}$, $I_{A=B}$, and $I_{A<B}$ inputs of the next higher stage. Stages can be added in this manner to any length, but there is a propagation delay of about 15ns added with each additional stage. For proper operation the expansion inputs for the

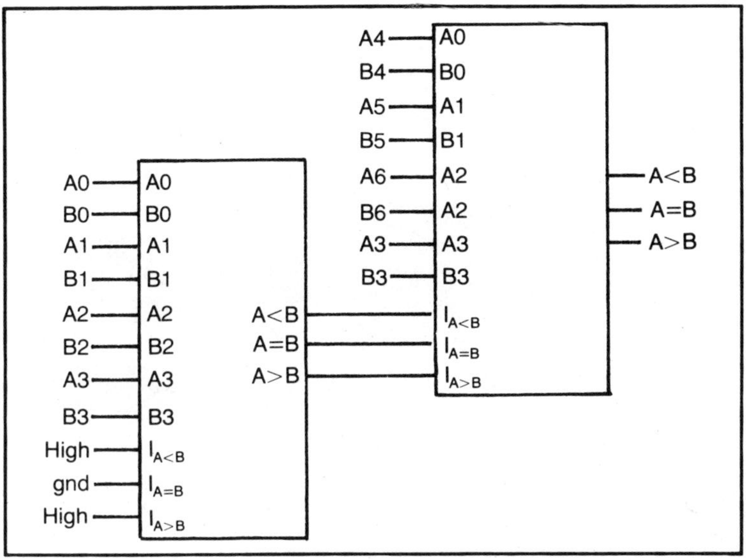

Fig. 6-28. Connecting two 7485 comparators in series to compare two 8-bit words.

least significant stage should be biased with $I_{A>B}$ high, $I_{A=B}$ low, and $I_{A<B}$ high.

Figure 6-28 shows a series connection of two 7485 comparators connected to compare two 8-bit computer words. These two comparators are connected in series with the A=B output only when all eight inputs are equal.

Due to the additional delay for the series connection, it is possible to create excessive and intolerable delays for larger number of bits. In this case, a parallel expansion scheme can be used. In this approach, one 7485 is used to collect the data from several comparators operating in parallel. The least significant comparator outputs are connected to the expansion inputs of the common 7485. The more significant comparator A>B and A<B outputs are connected to the A and B inputs of the common comparator. This method can be used to compare up to 24 bits.

7

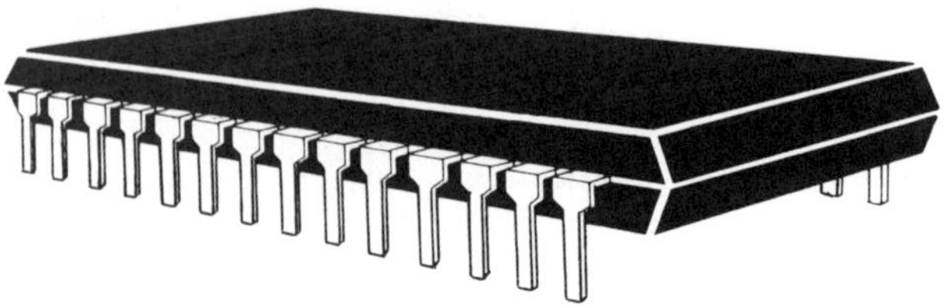

Analog Applications

Five different analog integrated circuits have been introduced. I discuss typical applications and application information in this chapter. The next chapter covers the actual use of the analog ICs in projects.

A typical analog integrated circuit can be used for several different applications. The applications are all related in some form to the basic application, so when the basic use of the IC is understood, the related applications are easier to understand. This information lays the groundwork for the projects in the next chapter.

Most analog integrated circuits can be powered by a wide range of voltages. The circuit may be powered from batteries, from an available power supply, or from the system power source, thus giving the builder more flexibility.

MC14066B ANALOG SWITCH

The MC14066B analog switch is an electronic equivalent of a single-pole single-throw switch. It is either on or off and is controlled by a control input. it is used in both analog and digital circuits, but it is classified as an analog IC because it does not operate under the principle of only two stable states for the output. There are only two states for the switch, on and off, but the output is either a high impedance, or a reproduction of the input. The input may be an analog signal with any value between full on and full off.

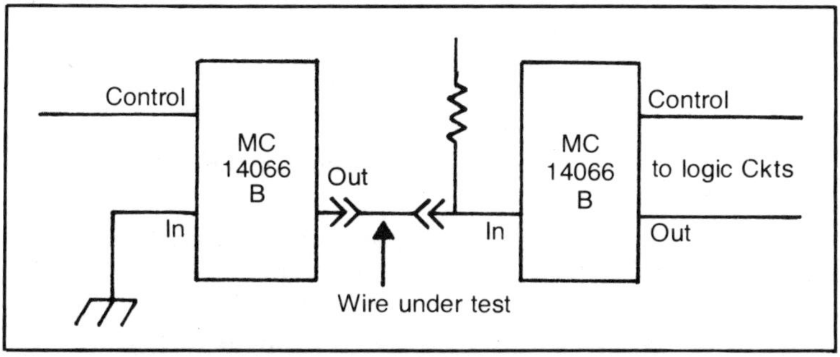

Fig. 7-1. Analog switch circuit for measuring continuity of a wire.

Automatic Testing

Figure 7-1 shows an application for the MC14066B analog switch. This circuit is one of four identical circuits to measure for continuity and shorts between the wires in a four wire cable. A ground is applied by the left switch, and the right switches are enabled one at a time. If the ground is seen by any circuit other than the wire that is grounded, this indicates a short. If the ground does not appear on the proper wire, this indicates either an open or wrong wiring.

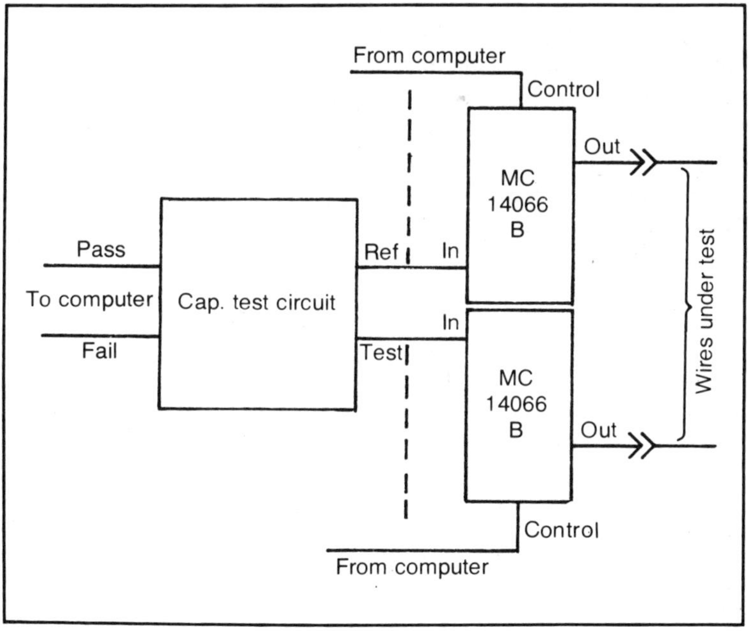

Fig. 7-2. Analog switch circuit for applying test signals to a pair of wires for measuring capacitance between the wires.

The control inputs are controlled and sequenced by a microcomputer, and the output of the right switch goes to logic circuits and is gated into the computer. Thus the computer sequences through the test until the opens and shorts are repaired. This application uses the analog switch to sequence through a part of an automatic test.

Figure 7-2 shows another application of the MC14066B analog switch in an automatic test set. This circuit uses the analog switches to connect two wires to the capacitance test circuit to measure the capacitance between the wires on a pass/fail type test. The computer controls sequencing with the analog switches. Each one is connected to different wires, in such a way that every wire can be connected to each of the two test signals. The use of analog switches with automatic test equipment permits easy control of the switches by the control computer. If sequencing and the test signal frequency is within the limits of the switch, the analog switch can be used to switch different test signals to different places in the unit under test.

Multiplexing

Another common application is shown in Fig. 7-3. This circuit is a simple multiplexing circuit, which allows the sampling of four input signals and combining them to form one output signal. This is commonly used in telemetry where several signals are transmitted over one line. Each input has its own time slot in the output, and the time pulse enables the switch to place the inputs on the output signal. The time pulses are sequential, and control the sampling rate.

The demultiplexing of the signal is done in much the same manner, except that a sync pulse is generated to synchronize the transmitter time pulses with the received time pulses. Other analog switches are available which have the outputs connected together, thus allowing easier application.

Drivers

Another application of analog switches is to drive load circuits from digital circuits. Since the through current for this switch is 25 mA maximum, this can be used as a current source for a 25 milliamp load. This is illustrated in Fig. 7-4. The analog switch is driven by a logic gate output. This overcomes the problem of driving loads which require current from the driving device.

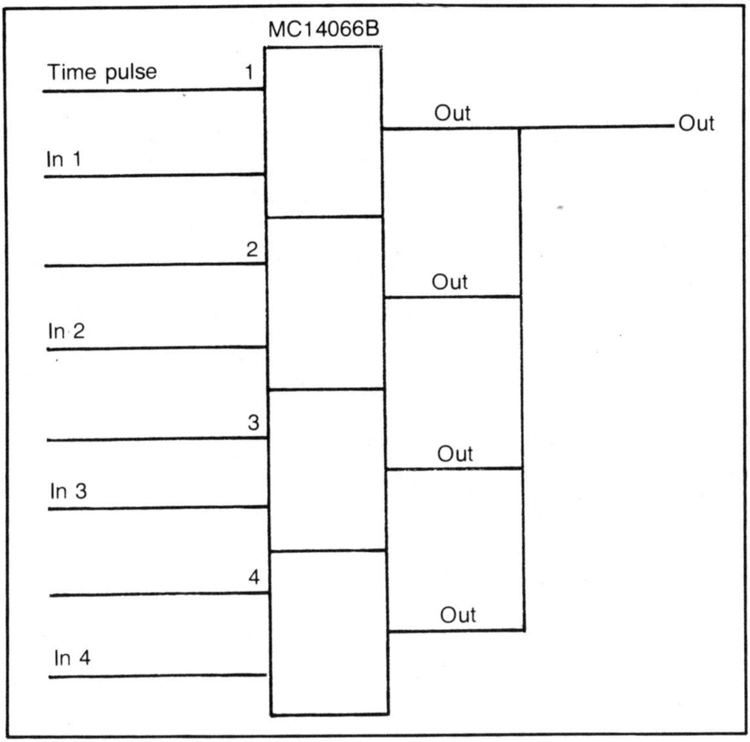

Fig. 7-3. Multiplexing four signals with an analog switch.

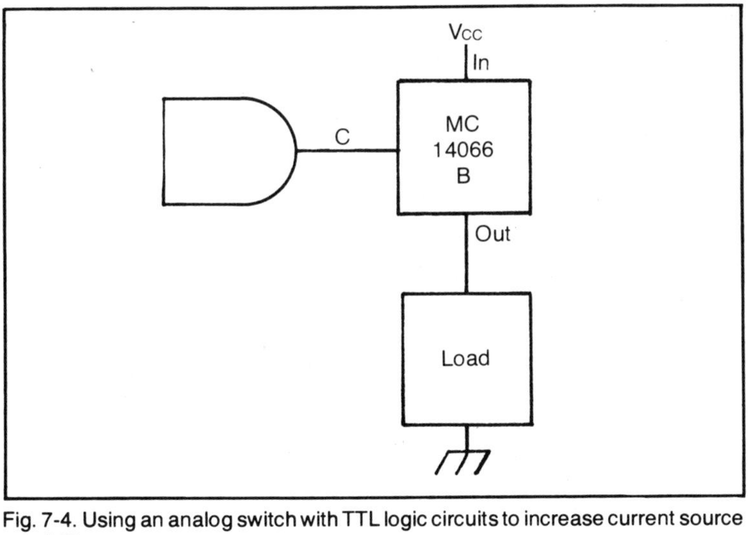

Fig. 7-4. Using an analog switch with TTL logic circuits to increase current source capability.

The 555 timer circuit is an IC which is designed for one shot and oscillator applications. It can also be used in related applications such as a pulse width modulator, pulse position modulator, or linear ramp generator. It uses external resistors and capacitors to establish the time of operation. The various functions are achieved by connecting the various input and control functions in different configurations. Refer to Chapter 5 for a detailed description of how the circuit operates.

As with most timing circuits, the 555 uses a resistor and a capacitor to determine the pulse width. The charge on a capacitor builds up as a function of time, as shown in Fig. 7-5. This is a simple RC circuit, and when the switch is closed, the voltage starts from zero and builds up in an exponential manner. This is shown by the waveshape of the capacitor voltage.

The rate of charging of the capacitor, and consequently the rate of change of the capacitor voltage, is determined by the values of the resistor and capacitor. The larger these values, the longer it takes for the voltage to build up to a predetermined level. If the threshold level is set at 2/3 of the full voltage, changing the values of the resistor and capacitor will change the time it takes for the capacitor voltage to reach this level. The exact equation is a function of what

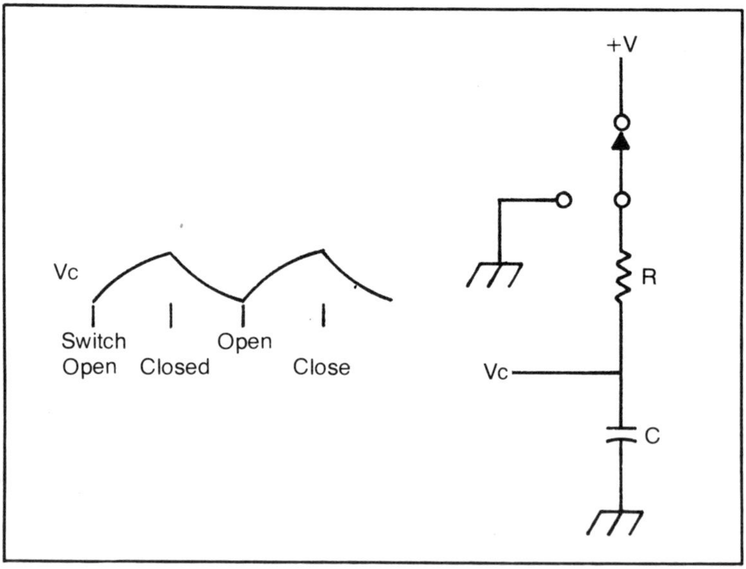

Fig. 7-5. Charging and discharging voltage of a resistance and capacitance circuit.

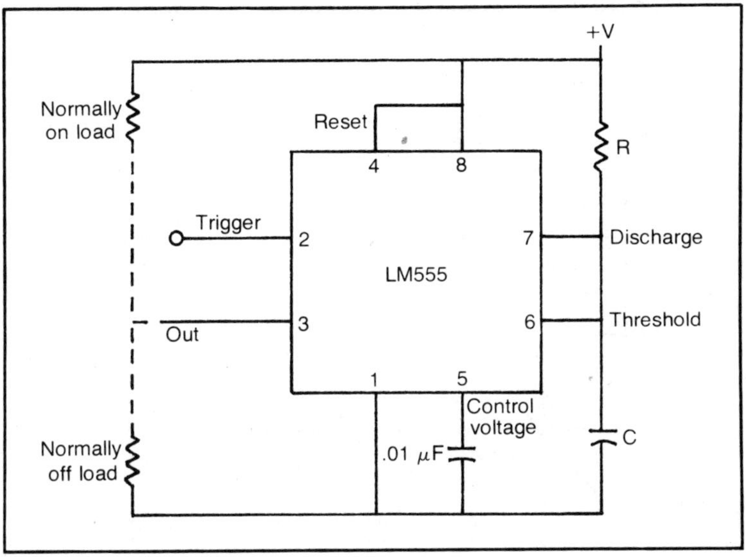

Fig. 7-6. The 555 timer used in a one-shot circuit.

is connected to the capacitor, and the current requirements of this load.

One Shot

The basic application for the 555 timer IC is as a one shot. The circuit has one stable condition, and the trigger input will force it into the unstable condition. It is held in the unstable condition until the circuit times out and the capacitor voltage has reached the threshold level. Then the circuit times out, returning to the stable condition. Figure 7-6 shows the 555 connected as a one shot. R and C are the timing components, and are defined by the following equation:

$$t_{sec} = 1.1 \times R \times C$$

R is in ohms and C is in farads. If R is 10 K ohms and C is 0.1 microfarad, then t is 1.1 milliseconds. Figure 7-7 gives the values of R and C against time. This chart can be used to obtain the values required for R and C for a given time. Due to the exact values of the resistor and capacitor, exact pulse widths are difficult to obtain. If the tolerances on the components are greater than the tolerance on the pulse width, a variable resistor may be used. This allows adjusting the resistor for the exact pulse width required.

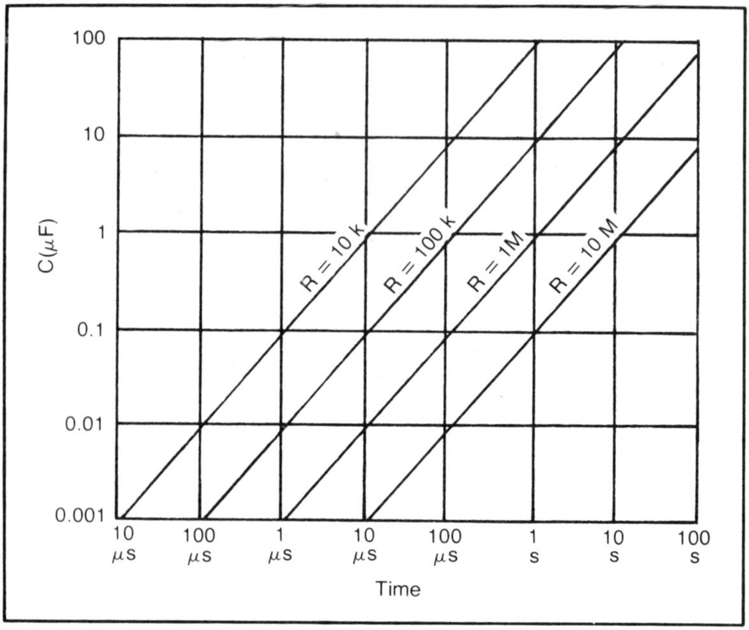

Fig. 7-7. Resistor and capacitor chart for one-shot pulse times.

The trigger input initiates the timing cycle. It must be a negative going pulse because the circuit triggers on the level change. The trigger level is about ⅓ of the applied voltage, so for a 5 volt power supply, the circuit will trigger at about 1.7 volts. This is compatible with TTL digital logic circuits. The 1 state is above this level and the 0 state is below this level.

When the timing cycle is terminated, the discharge terminal on the IC will discharge the capacitor in preparation for another trigger input. If the discharge rate is limited by a resistor, the capacitor voltage will decrease in an exponential manner. If the trigger input is connected to the capacitor, it will sample the capacitor voltage. It will trigger the circuit when the voltage falls below the trigger voltage. Time out will occur when the threshold voltage level is achieved. So the circuit will continue to change states, with the capacitor voltage fluctuating between the threshold level and the trigger level.

Oscillator

Figure 7-8 shows the 555 connected as an oscillator. R_B is the resistor which controls the discharge rate of the capacitor, but it also controls the charge rate. Since R_A also controls the charge rate, the charge time and the discharge time cannot be made equal with this circuit. The timing equations for the output high time and the output low times are:

$$t_{high} = 0.693\ (R_A + R_B) \times C$$
$$t_{low} = 0.693\ (R_B) \times C$$

So the total time for one period is:

$$t_{total} = t_{high} + t_{low} = 0.693\ (R_A + R_B) \times C$$

The frequency is $1/t_{total}$. The equation for the duty cycle is:

$$D = R_B/(R_A + 2R_B)$$

From this equation, it is apparent that the only way to set the duty cycle at 50 percent ($t_{high} = t_{low}$) is to let $R_A = 0$. But this connects the

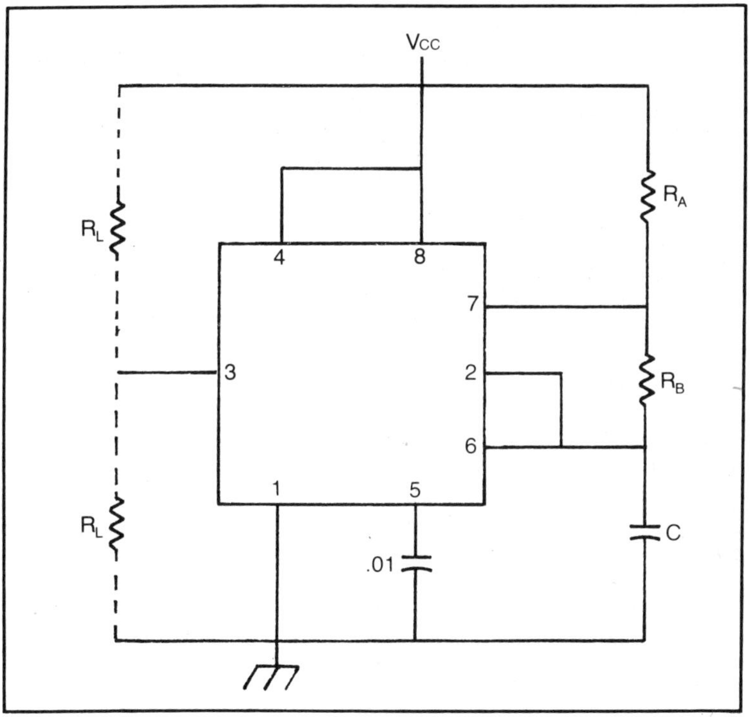

Fig. 7-8. The square wave generator circuit for the 555 timer.

discharge pin to +V, and provides for no current limiting. The current into this pin must be limited to a maximum of 200 mA and should be limited further to reduce the current requirements of the IC.

Figure 7-9 gives the timing chart for this circuit. The values of R given in the chart are the sum of $R_A + 2R_B$. After R is determined, the individual values must be determined from the duty cycle equation.

This makes a stable, useful square wave oscillator, requiring only four external components. There is a wide variation in the frequency attainable by changing the values of resistance and capacitance. For frequencies less that 1 Hz, the charge and discharge times of the timing circuit become long. For this condition, low leakage capacitors must be used. Otherwise, the capacitor leakage current can become a significant part of the discharge current. This affects the time of the oscillator.

Figure 7-10 shows the schematic for a wide range frequency-adjustable square wave oscillator that is capable of operating from +5 volts. If the upper variable resistor is set to zero, the 100 ohm

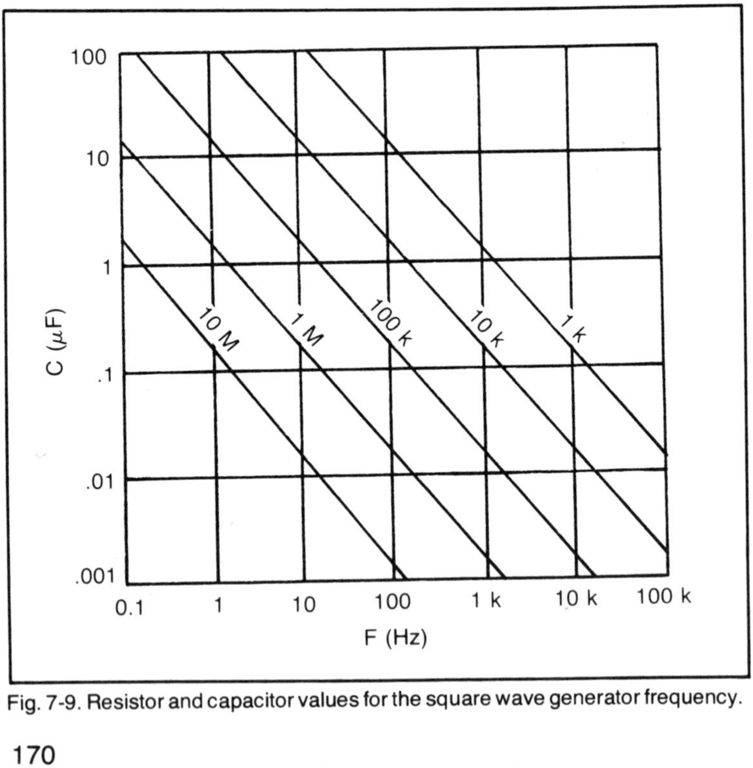

Fig. 7-9. Resistor and capacitor values for the square wave generator frequency.

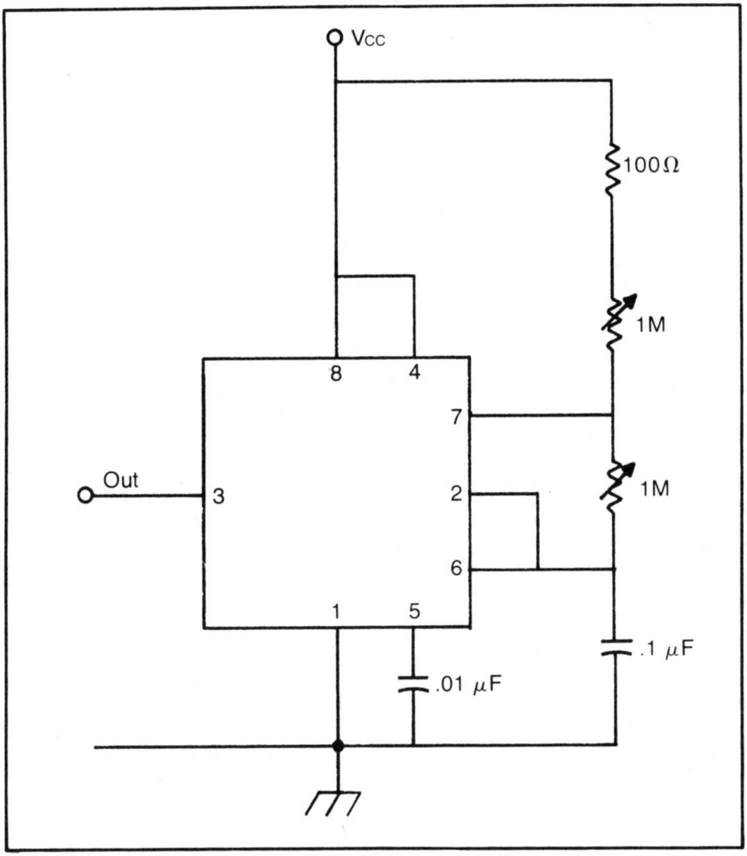

Fig. 7-10. An adjustable frequency square wave generator for the 555 timer.

resistor limits the current to the discharge pin to 50 milliamps, well within the limit. With both variable resistors set to maximum, the frequency will be less than 10 Hz, with a duty cycle of 33 percent. If the upper variable resistor is set to zero and the lower to 500 ohms, the frequency of operation becomes about 10 kHz.

Figure 7-11 shows the schematic for a circuit which can achieve a 50 percent duty cycle. The frequency of operation becomes a complex function of the resistors and capacitor, and is beyond the scope of this book. It is presented here only as an example.

Modulator

Using the control voltage input pin as a modulation input opens up

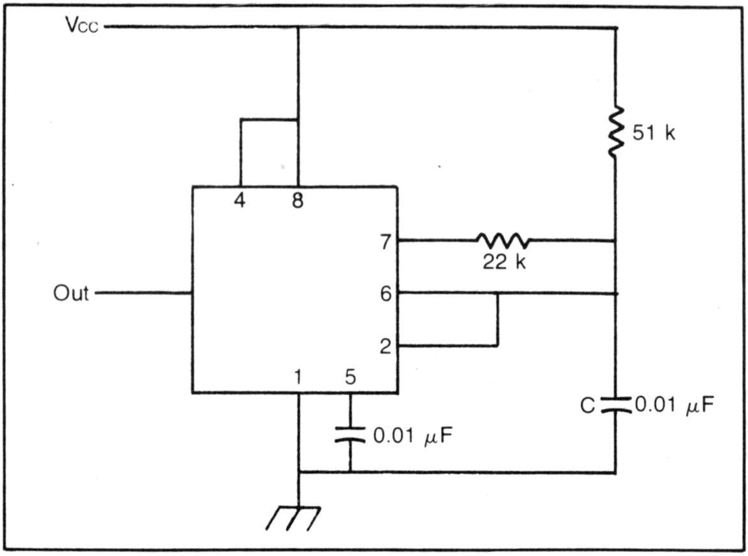

Fig. 7-11. A 555 timer circuit for a 50 percent duty cycle square wave generator.

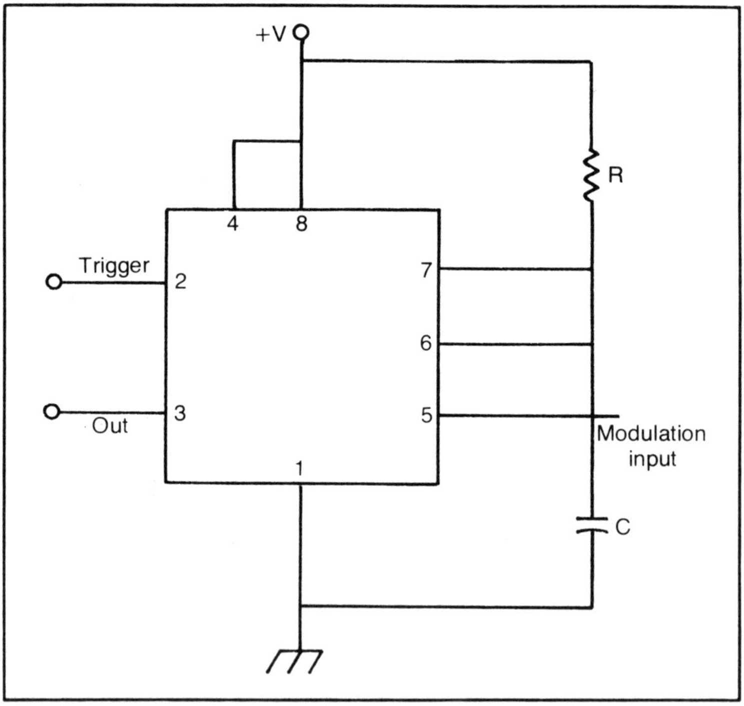

Fig. 7-12. A 555 timer pulse width modulator circuit.

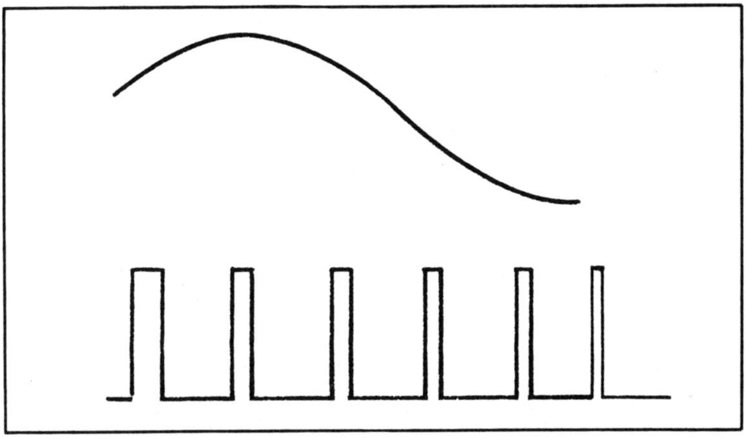

Fig. 7-13. Waveshapes for Fig. 7-12.

two interesting applications. The first of which is a pulse width modulator, as shown in Fig. 7-12. A pulse width modulator generates a series of pulses whose width is dependent on the amplitude of the modulating signal. The times the pulses occur is determined by

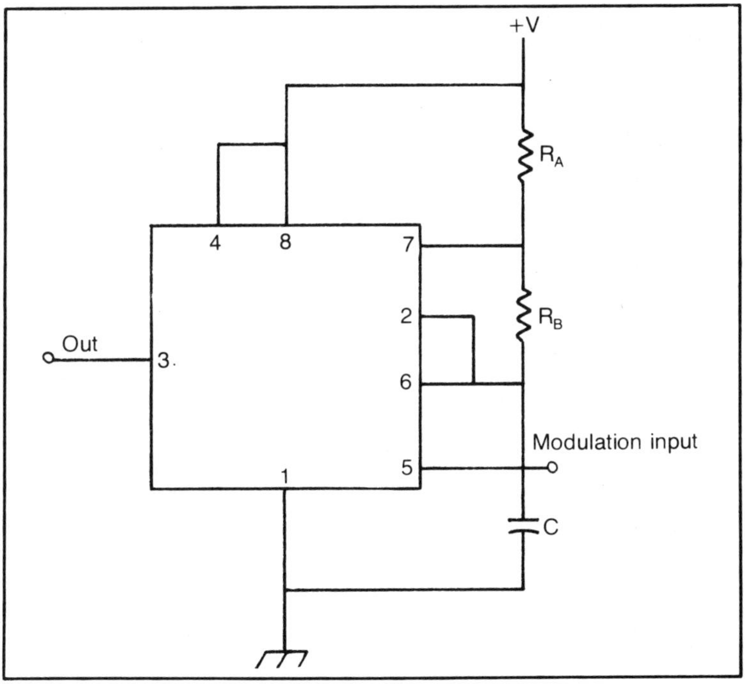

Fig. 7-14. A pulse position modulator using a 555 timer.

the trigger input which triggers the one shot. The pulse is terminated when the modulation input equals the threshold, so it is a function of the amplitude of the modulation input. Figure 7-13 shows the waveshapes for the modulating input and the output.

Another application using the control voltage input is the pulse position modulator, as shown in Fig. 6-14. This circuit generates pulses whose width is a function of the modulating input, and the time between the pulses is fixed. This is basically a free-running square wave generator with its high time as a function of the modulation input, and its low time fixed. Figure 7-15 shows the associated waveshapes.

Since the timing functions are based on comparing one voltage level against another and since the voltage levels are a function of the applied voltage, the exact value of the applied voltage is not critical. The applied voltage can change from 5 volts to 15 volts with a very little change in the times, although the current drawn by the circuit will increase. The main change will be in the trigger level required.

LM340T-5 VOLTAGE REGULATOR

Three terminal voltage regulators have greatly simplified power supply design. They replace the regulator section of a power supply and provide additional capabilities. One three terminal IC replaces several resistors and capacitors and a few transistors. Figure 7-16 shows the circuit for a fixed output regulator. The output voltage is determined by the design of the IC itself and is reflected in the part number. For the LM340T series of regulators, there are eight

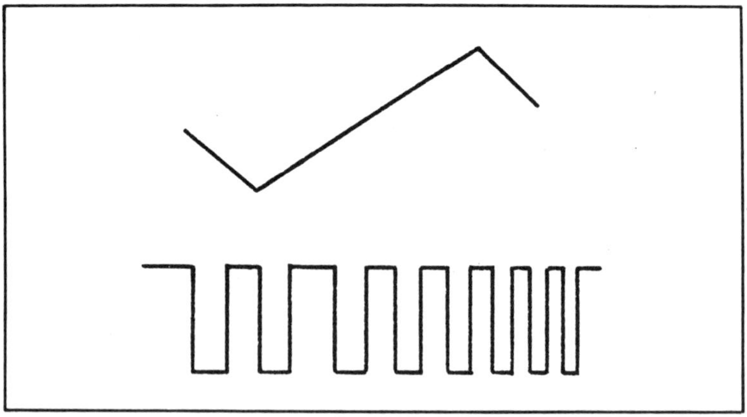

Fig. 7-15. Waveshapes for Fig. 7-14.

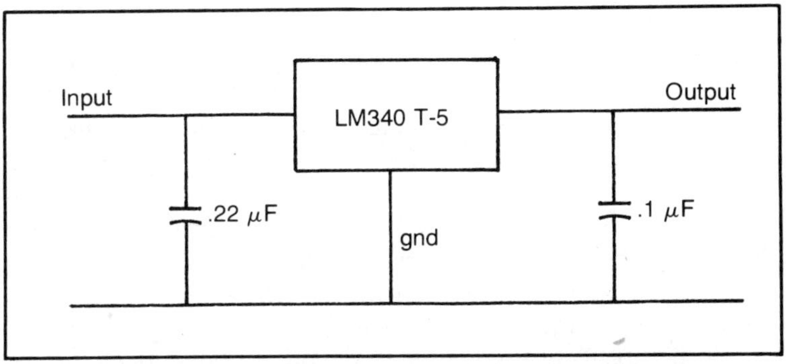

Fig. 7-16. Typical three-terminal voltage regulator fixed output circuit.

different positive output voltages available. The application is similar for all the different dash numbers, and the 5 volt regulator was chosen because 5 volts is required for the TTL logic circuits.

Returning to Fig. 7-16, the input voltage is usually the voltage from the rectifier portion of the power supply. It must be at least the minimum voltage required to maintain regulation. The input capacitor is required if the regulator is not physically a part of the rectifier portion, and is located remotely. This capacitor provides filtering and transient suppression on the input voltage. The output capacitor improves the transient response of the regulator. For some voltage regulators, the output capacitor is required for stability, but for the LM340 series it is not required. It is a good practice to include it, however, to improve the transient response of the regulator.

Figure 7-17 shows a simple bridge rectifier power supply. This

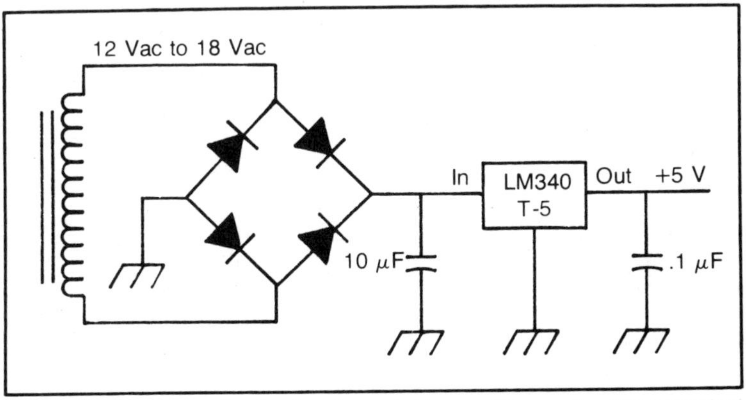

Fig. 7-17. Power supply circuit using a bridge rectifier and a three-terminal voltage regulator.

circuit can use a wide variety of transformers to provide the low voltage ac for the rectifiers. The bridge rectifier provides full-wave rectification of the input voltage, and the 10 μF capacitor provides smoothing of the full-wave voltage. This voltage is fed into the regulator which controls it to 5 volts for the output voltage. This type of power supply is used by many hobbiests for powering circuits. If the input voltage to the regulator is higher than required, the voltage drop across the regulator will increase and the power dissipated will increase. As long as the current required does not cause thermal shutdown, there is no problem.

If additional current is required, more than one regulator may be powered from the same rectifier circuit providing the loads are separate. For instance, some of the ICs in the circuit can be powered by one regulator while other ICs are powered by another regulator. The outputs of the regulators cannot be connected together because the exact voltage out will result in one regulator providing most of the current. If the load is common, and cannot be separated, it is possible to increase the current capability by connecting a power transistor to the output and using the regulator to establish the reference voltage. The circuit for accomplishing this is complex, and the equations for determining the component values is complicated.

If there is no regulator available for the next voltage required, some things can be done to increase the voltage at the regulator output. These involve raising of the ground pin of the regulator to a voltage above ground. Figure 7-18 shows one method of accomplishing this using the forward voltage drop across a diode. The current through the diode is the regulator quiescent current so the diode voltage drop must be fairly stable and beyond the knee of the diode curve. If this is not practical or possible, it is possible to bias the diode by connecting a resistor from the top of the diode to the input voltage, as shown in Fig. 7-19. This assures that the voltage is

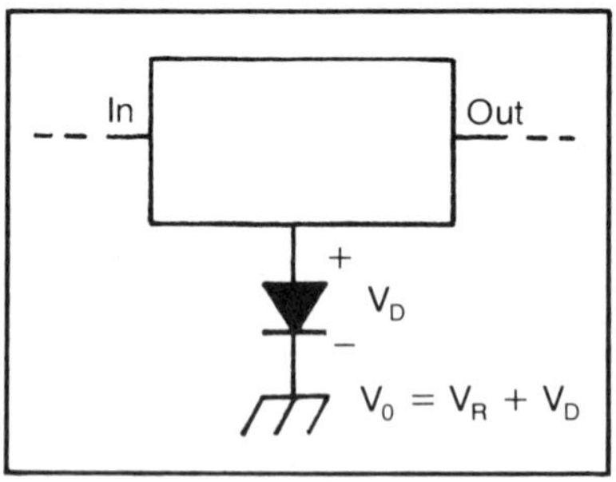

Fig. 7-18. Increasing the output voltage by using a diode in the common line.

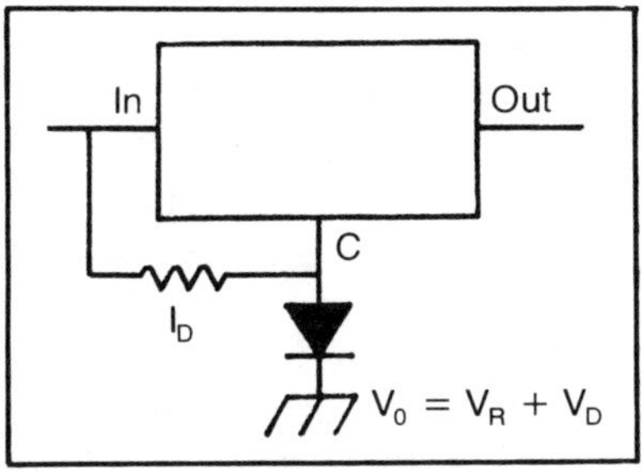

Fig. 7-19. Current biasing the diode of Fig.7-18 to assure the diode is operating beyond the knee of the diode curve.

stable and that the diode is beyond the knee of the curve. More than one diode can be used, if required. The output voltage is the regulator voltage plus the diode voltage drops.

Another method of increasing the output voltage is to connect the regulator as shown in Fig. 7-20. This circuit makes an adjustable regulator where the output voltage is the regulator voltage plus the voltage drop across the variable resistor. The output voltage is defined by the following equation.

$$V_{out} = V_{reg} + V_{R2} = 5\text{ V} + (I_Q + 5\text{ V}/R1)R2$$

I_Q is the quiescent current.

The value of resistor R1 must be such that 5 volts divided by R1 is greater than 3 times the quiescent current. The load regulation is degraded by a factor of (R1+ R2)/R2 and becomes:

$$R_{load} = \left(\frac{R1 + R2}{R2} \right) \times R_{load\ spec}$$

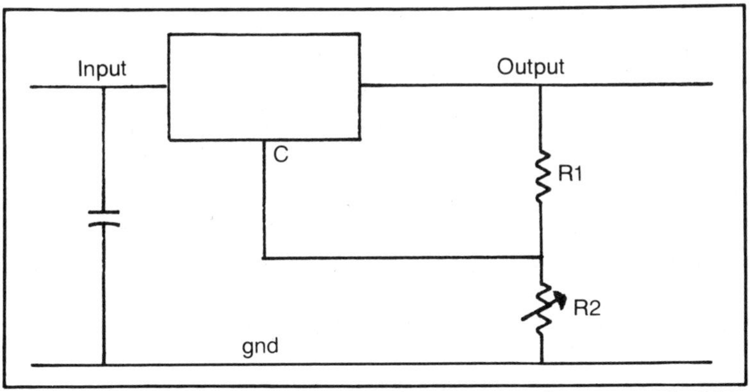

Fig. 7-20. Using the LM340T-5 three-terminal voltage regulator as an adjustable voltage regulator.

Where $R_{load\ spec}$ is the load regulation from the specification sheet. The value of R2 with respect to R1 is determined by the voltage adjustment range desired. the ratio can be calculated from the output voltage equation given previously. When R2 is set to zero, the output voltage becomes 5 volts. This is the minimum adjustment.

Another type regulator sometimes required is the current regulator. This regulator, as shown in Fig. 7-21, controls the current by changing the voltage drop across the series resistor, R. The output current is given by the following equation:

$$I_{out} = \left(\frac{V_{out-c}}{R1} \right) + I_Q$$

V_{out-c} is the voltage between the output of the regulator and the common terminal. This is the specified output voltage of the regulator.

LM324 OPERATIONAL AMPLIFIER

As discussed previously, the operational amplifier is one of the most versatile of the types of analog circuits available. It has found many applications, from the simple amplifier circuit, to oscillators, active filters, integrators, and analog computers. It can be used almost anyplace that the desired function can be represented by an equation that can be represented by electronic components.

The ideal amplifier has infinite input impedance, zero output impedance, infinite open-loop gain, infinite frequncy response, and

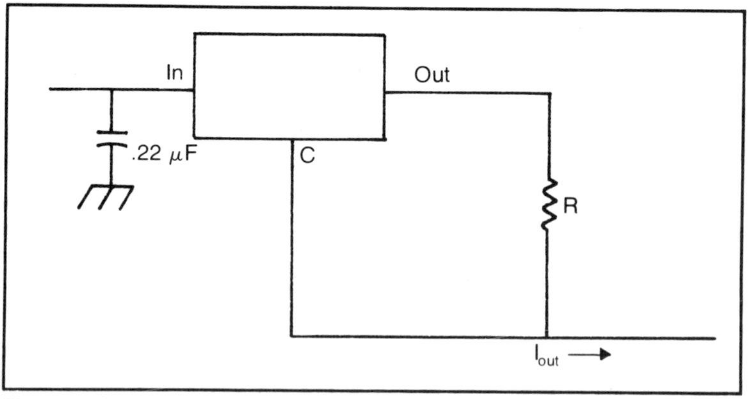

Fig. 7-21. Constant current regulator.

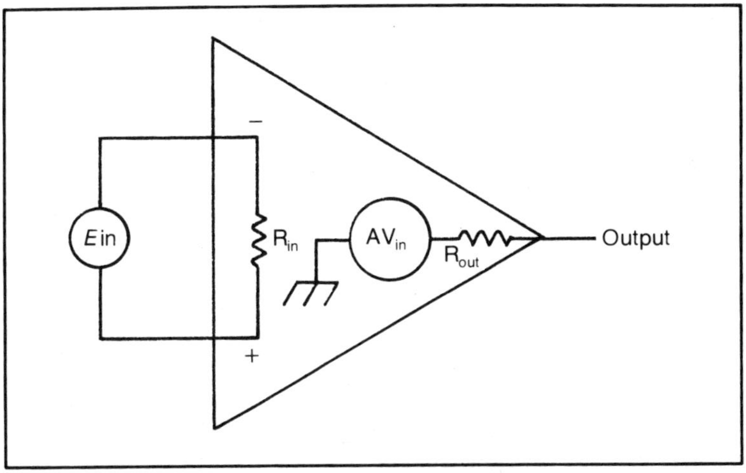

Fig. 7-22. Equivalent circuit for an operational amplifier.

zero offset. This leads to two assumptions or approximations used when working with operational amplifiers.

- ☐ No current flows into either input terminal.
- ☐ The voltage between the two input terminals is zero. That is, the differential input voltage is zero.

These are valid assumptions for the first order analysis and for the design of circuits well within the specifications of the amplifier.

The equivalent representation of an operational amplifier is shown in Fig. 7-22. It is represented by an input resistance, an output resistance, and a signal generator whose output is equal to the gain times the input voltage. This equivalent circuit is used for rigorous analysis, which is beyond this book.

Inverting Amplifier

Since the amplifier has a very large gain, stability is an important design consideration. A small signal will drive the circuit full on or full off, and since the current required for the drive is small, the signal could be induced electrical noise. To stabilize the amplifier and make the operation predictable, feedback from the output to the input is used. A simple amplifier, with feedback, is shown in Fig. 7-23. This is an inverting amplifier, where the output is inverted from the input signal. The two assumptions are used in determining the performance equations for this circuit.

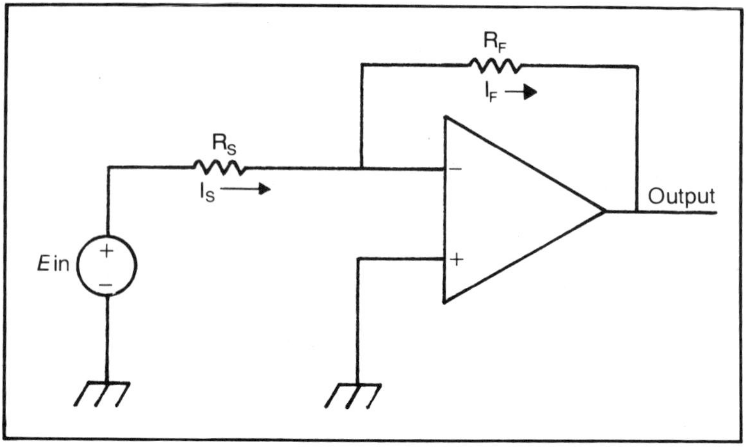

Fig. 7-23. The basic inverting amplifier circuit.

Since the voltage at the inverting (−) input equals the voltage at the noninverting (+) input, the currents are defined in the following manner:

$$I_s = E_{in}/R_s$$
$$I_f = E_{out}/R_f$$

Since it is assumed that no current flows into the amplifier, $I_s = I_f$. Therefore the voltage gain is defined this way:

$$A = E_{out}/E_{in} = -R_f/R_s$$

The minus sign for gain indicates that the output signal is inverted from the input signal. The input impedance of the amplifier is R_s and the output impedance is zero.

The operation of this operational amplifier circuit can be visualized as feeding back a portion of the output signal into the minus input so as to control the gain. The feedback is such that the gain of the operational amplifier itself no longer enters into the circuit gain equation. The resistor combination determines the gain, and gain is equal to the feedback resistor divided by the series resistor.

Summing Amplifier

Because of the virtual ground at the minus input to the amplifier, any number of input voltages may be summed through individual series resistors at the minus input of the op amp. Each input signal operates through its respective input resistor, and the current in the feedback resistor is the algebraic sum of the current from each input source. This makes a summing amplifier, as shown in Fig. 7-24. If all the resistors are equal, the output voltage becomes:

$$V_{out} = e_1 + e_2 + e_3$$

If the resistors are not equal, then the output voltage is dependent on the gain for each series resistor.

$$V_{out} = (e_1 \times R_f/R_{s1}) + (e_2 \times R_f/R_{s2}) + (e_3 \times R_f/R_{s3})$$

If the series resistors are made equal, the summing amplifier will have gain. If the series resistors are not equal, each inputs can be weighted with a different gain.

Noninverting Amplifier

If the input to the basic amplifier circuit is connected to the plus (+) input, as shown in Fig. 7-25, the result is a noninverting amplifier. Since no current flows into the amplifier inputs, R_f and R_s form a simple voltage divider connected to the minus input. The voltage at this input becomes:

$$E_i = \left(\frac{R_s}{R_f + R_s} \right) \times E_{out}$$

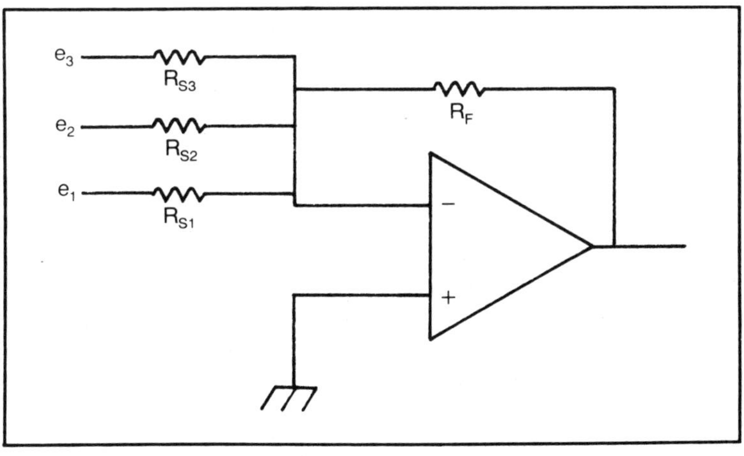

Fig. 7-24. An operational amplifier summing amplifier.

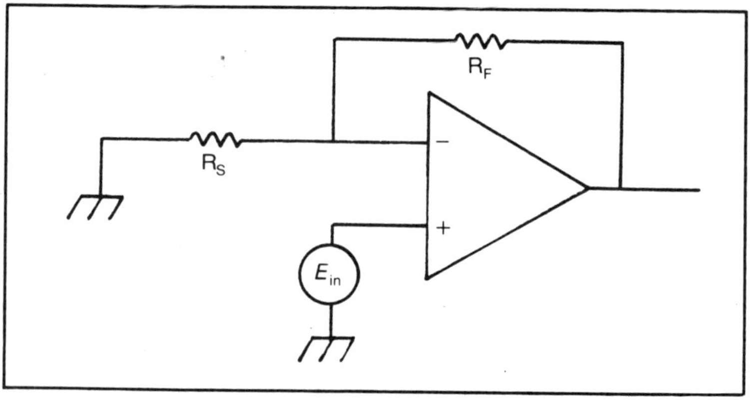

Fig. 7-25. The basic noninverting operational amplifier circuit.

This voltage must be equal to the input voltage because of the infinite gain, and, therefore, the gain of the circuit is:

$$A = \frac{E_{out}}{E_{in}} = 1 + R_f/R_s$$

In this case, the input impedance approaches infinity, and the output impedance approaches zero. The minimum voltage gain for this circuit is one, or unity, and is achieved when R_f is made zero. If R_f is zero, then the second term in the equation is zero. A is then equal to 1.

Voltage Follower

Figure 7-26 is a voltage follower circuit where the gain is one and there are no feedback resistors. The output equation for this circuit is simply:

$$V_{out} = V_{in}$$

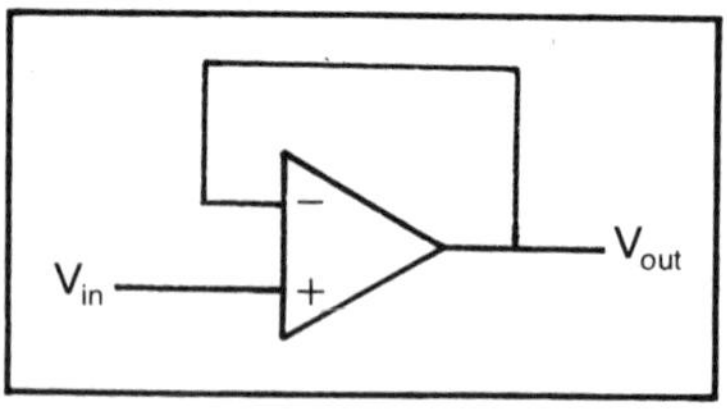

Fig. 7-26. Operational amplifier voltage follower circuit.

This circuit is used for isolation of the input circuit and has the highest input and lowest output impedance of the typical operational amplifier circuits. Input impedances above 10 megohms and output impedances below 1 ohm are readily obtainable with this circuit. This circuit is often used for isolation of sensitive circuits, or to reduce the loading of a signal by providing a buffer.

Differential Amplifier

Figure 7-27 shows the circuit for a differential amplifier. This circuit amplifies the difference between two voltages and generates a signal proportional to this difference voltage. The output voltage for this differential amplifier is:

$$E_{out} = \left(\frac{R_f}{R_s} \right) \times (E_{in2} - E_{in1})$$

The input impedance for this circuit is high, and the sensitivity to low level signals is good.

Current-to-Voltage Converter

Many transducers generate current as the output, so this current must be converted to voltage to be useful. The circuit in Fig. 7-28 is a simple current-to-voltage converter. The output voltage is given by the equation:

$$E_{out} = - I_{in} \times R_f$$

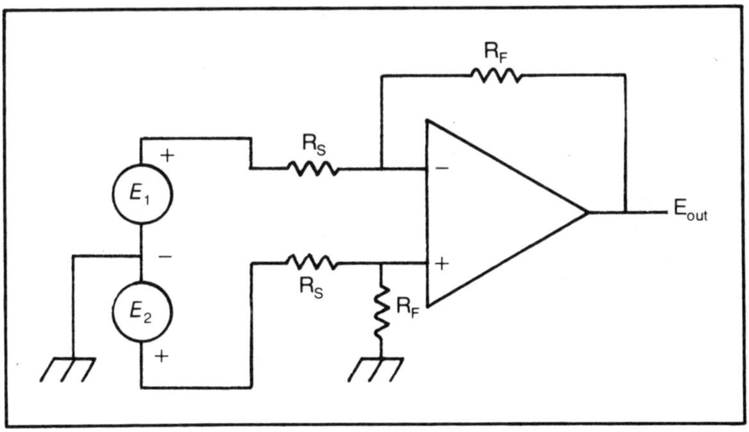

Fig. 7-27. Using an operational amplifier for a differential amplifier.

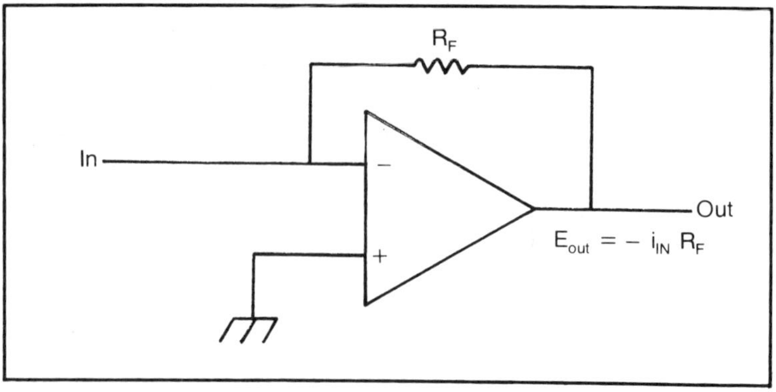

Fig. 7-28. An operational amplifier current-to-voltage converter.

So the output voltage is a function of the input current and the feedback resistor. This is basically a *zero drop* current shunt, with no voltage drop across the shunt. It provides an inexpensive method of measuring current. Not only is a voltage generated, but the current capability of that voltage is increased because of the low output impedance.

Bridge Amplifier

Many measuring devices use a bridge to balance the signal, and read the unbalance in the bridge as the output voltage. In a typical bridge, three arms are fixed and the fourth allowed to vary depending on the parameter being measured. A reference voltage is applied to opposite points on the bridge, and the voltage at the remaining two points is measured as the output. Any change in the variable arm of the bridge shows up as a change in the output.

Figure 7-29 shows a bridge amplifier. For the bridge, if all four resistors (R) are equal, the bridge is balanced and the output is zero. Any change in the variable resistor (the lower left resistor in the bridge) will change the voltage to the minus input of the amplifier. This is amplified as the output voltage. For the variable resistor, δ is the change in the resistance as a fraction of R. If this change is much less than 1 and R_f is much greater than R, then the output voltage is given by the equation:

$$V_{out} = V_{ref} \times \left(\frac{\delta}{2}\right) \times \left(\frac{R_f}{R}\right)$$

Variable Gain Amplifier

The gain of this circuit is set by the feedback resistor to the desired

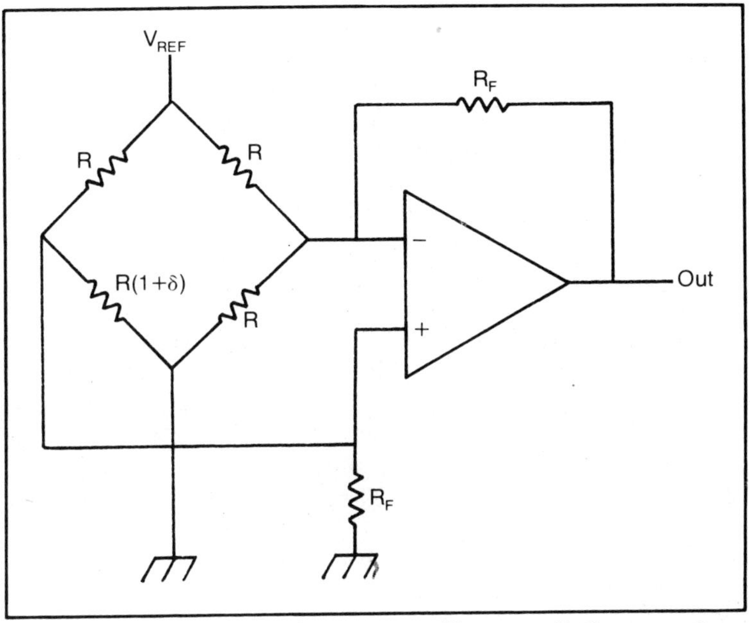

Fig. 7-29. An operational amplifier bridge amplifier to amplify the error voltage from a bridge circuit.

value for the error signal. If this circuit is used as a null amplifier so that the bridge can be nulled to zero, high gain must be provided.

Figure 7-30 shows a circuit with selectable gain settings. The switch is used to switch in different feedback resistors. An inverting

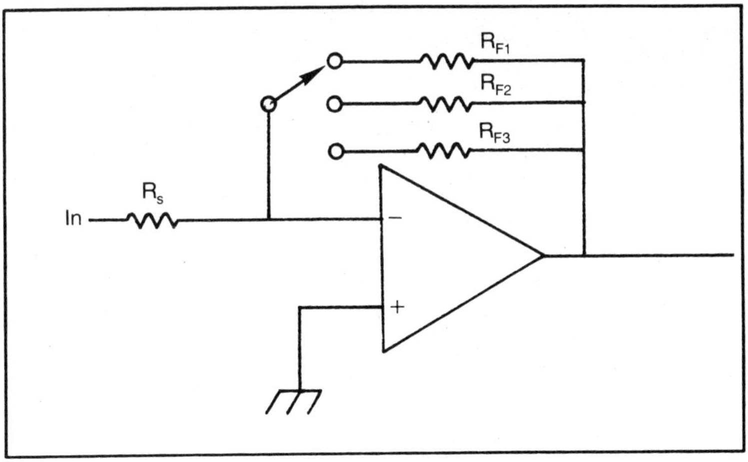

Fig. 7-30. Changing operational amplifier gain using a switch.

amplifier is used because the gain equation is straighforward. Changing the feedback resistor by a factor of 10 will change the gain of the circuit by a factor of 10. For instance if the series resistor, R_s, is 10 kilohms, and the feedback resistors are 10 megohms, 1 megohm, and 100 kilohms, the gains for the circuit are 1000, 100, and 10.

This type of circuit is often used as a null amplifier for a null bridge circuit. For a null amplifier, the object is to provide increasing gain for the error signal as it approaches zero so that the bridge can be precisely balanced.

An analog switch can be used to select the feedback resistors electronically. This is shown in Fig. 7-31. The value of the feedback resistance for the gain equation is the feedback resistor value added to the series on-resistance of the analog switch. So, for this application the feedback resistors are usually made variable resistors, and are adjusted for the gain desired. The control of the analog switches can be by analog circuits or by digital logic circuits.

An analog switch can also be used to sequence various signals

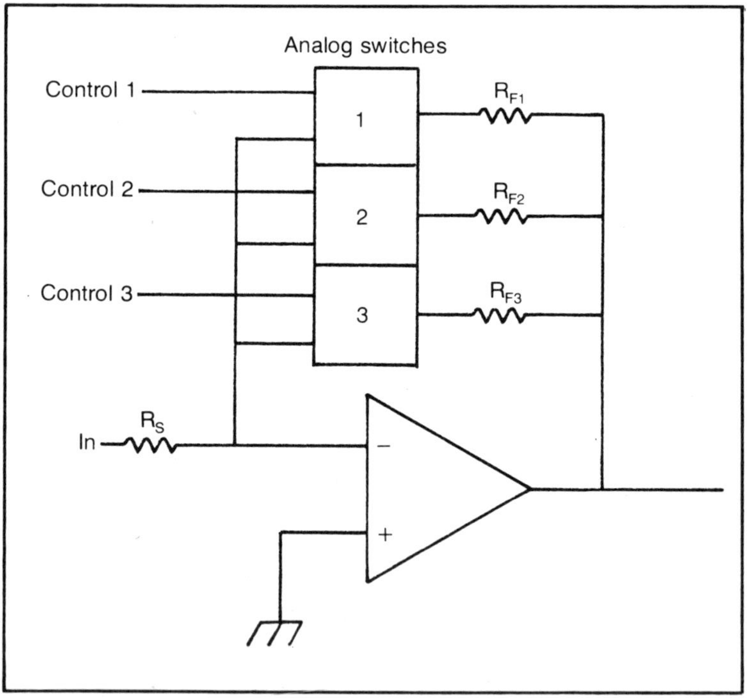

Fig. 7-31. Changing operational amplifier gain using an analog switch to select feedback resistor.

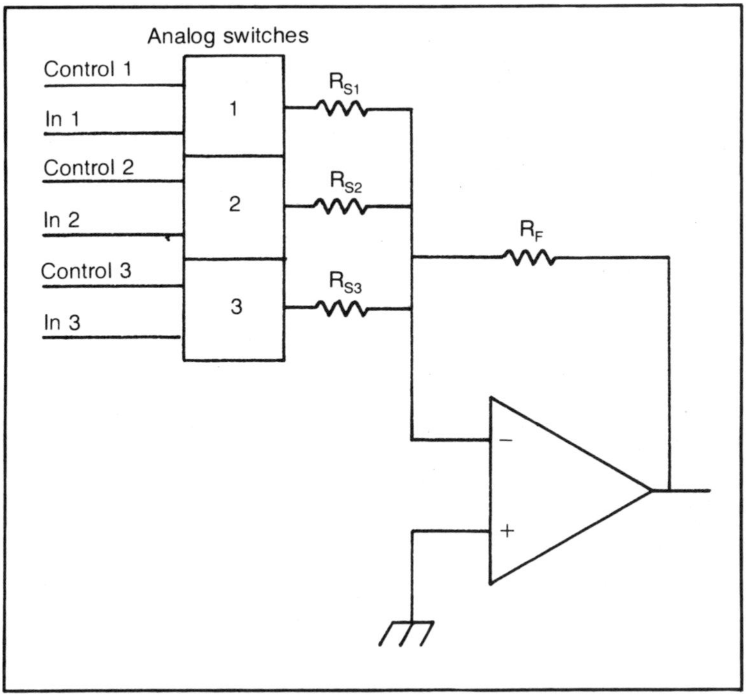

Fig. 7-32. Selecting the input signal to the operational amplifier circuit using an analog switch.

into one operational amplifier by placing the switch on the input, as shown in Fig. 7-32. The series resistor for the various input signals can be different for the different signals. This provides a means of providing different gains for the different inputs. Again, the value of the series resistor and the on-resistance of the analog switch must both be used in the gain calculation.

Comparator

A comparator circuit is a circuit which switches states when the input voltage reaches the level of the reference input. Such a comparator circuit is shown in Fig. 7-33. The resistors are connected to the positive input. This is to provide the circuit with an extremely high gain, and some hysteresis. The hysteresis separates the turn on point and the turn off point by a small voltage. This prevents rapid switching on and off of the circuit if the input voltage is slowly varying around the switching point.

The switching point is when V_{in} is equal to V_{ref}. If V_{in} is below

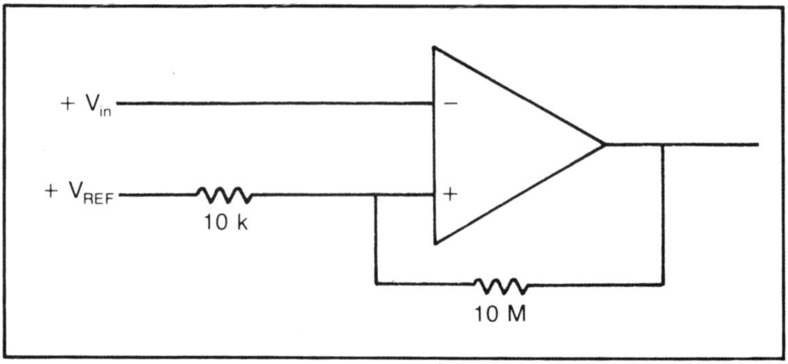

Fig. 7-33. Comparator circuit with hysteresis.

V_{ref}, there is a positive voltage difference at the input, so the output is high. If V_{in} is above V_{ref} there is a negative voltage difference at the input, so the output voltage is zero. Using feedback to the positive input terminal provides only two stable states for the amplifier, full on and full off.

Lamp Driver

If more current is required from the circuit than the operational amplifier can supply, a transistor can be used as a high current device. Such a circuit is shown in Fig. 7-34. This is a lamp driver, and depending on the transistor used, can drive up to 600 mA. The series and feedback resistors are not shown.

Design Considerations

There are some items which must be considered when building operational amplifier circuits. It is these points which can make the difference between a working circuit and a circuit which is acting

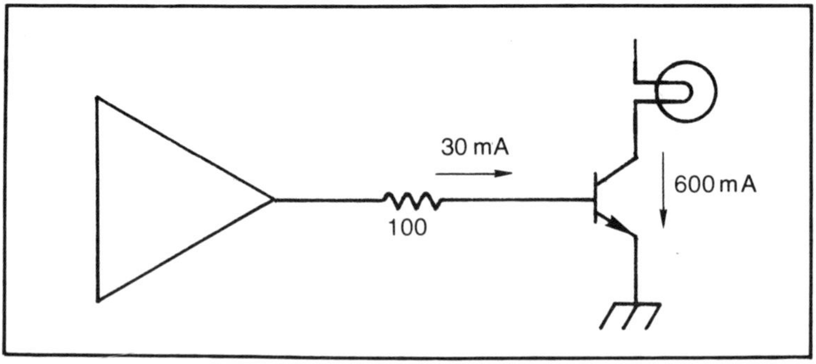

Fig. 7-34. Using a transistor to drive a high current load.

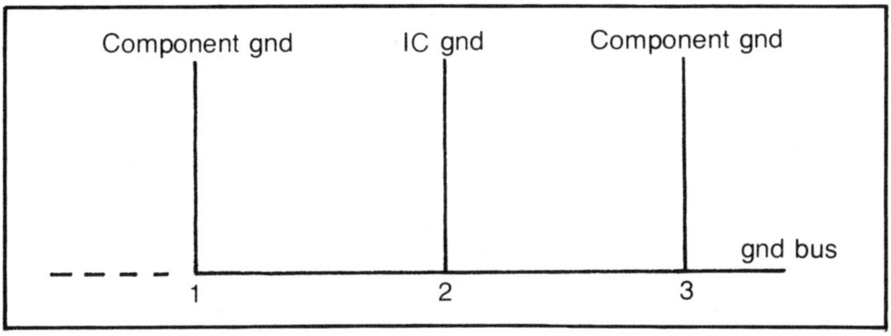

Fig. 7-35. Ground bussing.

strangely. The first point is the grounding of the circuit components and the integrated circuits which make up the circuit. The ideal ground is one in which all components are connected to the same point. This is called a single point ground. It means that the physical ground point is common to all input, power, output, bias, and IC grounds. Grounds should not be bussed, as shown in Fig. 7-35, for operational amplifier circuits. Since all conductors have a small, but real, resistance, there is a small voltage drop along the ground line.

If the ground current is flowing from 3 to 2 to 1 and on to the system ground, the current from 2 to 1 will generate a voltage drop along the ground line. If the current is varying, the voltage drop along the line will change, changing the ground reference for number 3. Also, any current from number 3 will cause point 2 to change in voltage. In other words, the voltage at point 2 will be the resistance between 2 and 1, times the current from 2 and the current from 3.

When the ground is varying in voltage, this can appear as a change in the signal level, and amplified as such. This consideration is especially important for low level analog signals where the change in the voltage at the ground may appear as an input signal, distorting or masking out the true input signal. The object of any good grounding scheme is to reduce this to a minimum. This is done by connecting each of the circuit grounds directly to the common ground point, as illustrated in Fig. 7-36. For this type of connection, the voltage at the ground at point 1 is a result of the current from point 1. One ground line does not affect the voltage at the other ground points. This is critical for low level amplifiers.

One of the assumptions made in the ideal circuit, is that they are stable. That is, they do not oscillate or exhibit ringing in response to the input signal, or in response to transients. The feedback must be negative from the output to the minus input to

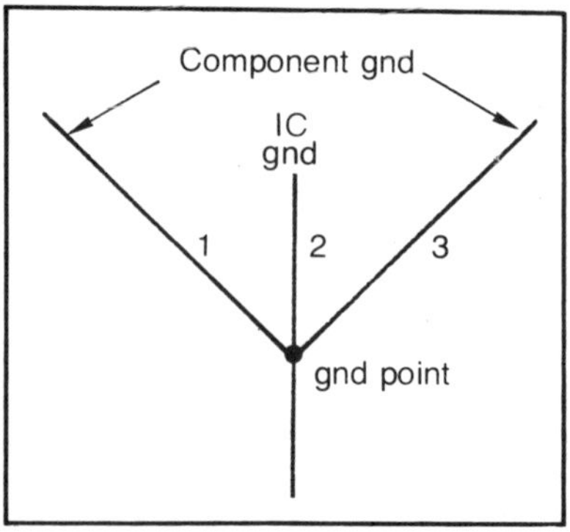

Fig. 7-36. The single point ground scheme.

maintain stability. The ideal operational amplifier has 180 degrees phase shift between the minus input and the output, generating the negative feedback. But the phase shift of the amplifier when combined with the feedback network may not be 180 degrees for all frequencies.

If there is enough phase shift in the feedback circuit to result in some areas of positive feedback for the input, instability can occur. This is because positive feedback amplifies the signal and results in a higher output voltage. This higher voltage, at the same frequency, is returned to the input by the positive feedback, where it is again amplified. The result can be oscillation at a significant output voltage level, and the input signal is lost. If the frequency for the positive feedback is high enough, the gain of the operational amplifier for that frequency may be less than one. If this is the case, there will be no positive feedback and no amplification of the feedback signal. But, if the frequency is such that the gain of the operational amplifier is greater than one, the circuit could be useless.

For a normal, resistive circuit, this phase shift is caused by stray capacitance due to wiring, leads, and the capacitive component of any circuit element. This shows up as shunt capacitance at the input summing point, and is illustrated in Fig. 7-37. C_s is the equivalent of the stray capacitance and may be 20 pF or more. To compensate for this, a compensation capacitor is used. One compensation method places the capacitor across the feedback resistor, as shown. The value of the capacitor is given by the following equation:

$$C_c = \frac{R_s \times C_s}{R_f}$$

C_s is the shunt capacitance. If there is a question about the value to use for C_s, realize that 20 pF is for a good printed circuit board layout, and the typical values will run many times this. This capacitor gives the starting point for determining the compensation capacitor. If fast response and a broad bandwidth is required, the compensation capacitor must be small, 10 pF to 100 pF. If fast response or high frequency operation is not required, higher values can be used.

Since the value of the shunt capacitance is difficult to determine, one method used to determine the size of the compensation capacitor is empirically. Operate the circuit at the maximum frequency required for the application, and try progressively larger capacitors. Keep this up until the output is affected, then use a capacitor about one half that value.

A properly constructed operational amplifier circuit can perform many functions, while a poorly constructed one can be a headache. Realize that most schematics given as typical applications do not include the compensation capacitor.

MC1306P AUDIO AMPLIFIER

The MC1306P is a ½ watt audio amplifier containing both the preamplifier and power amplifier. It provides the builder with a means of controlling preamplifier gain, power amplifier gain, input impedance, and frequency response. Each amplifier, the preamplifier and the power amplifier can be considered as special

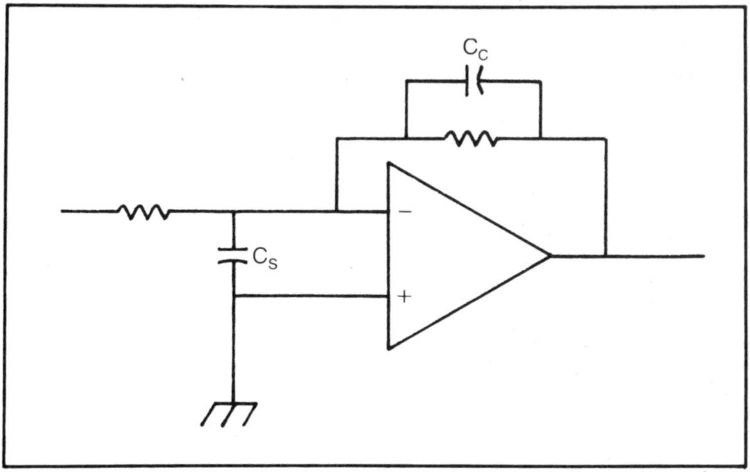

Fig. 7-37. Illustrating the shunt capacitance and compensation capacitance for an operational amplifier circuit.

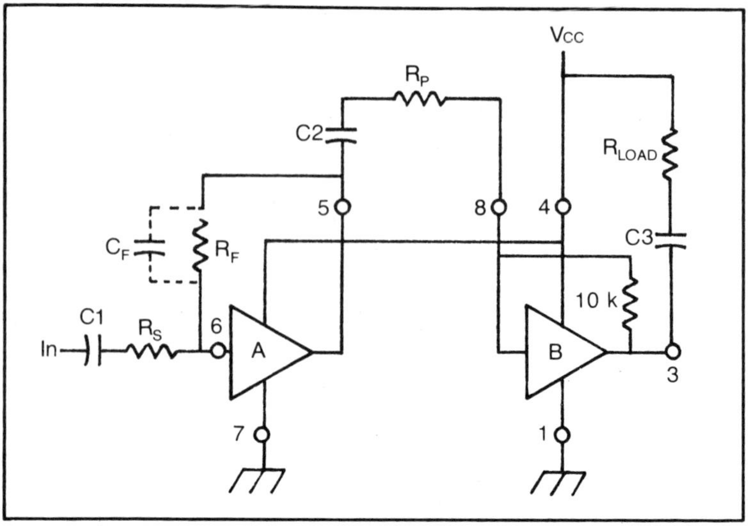

Fig. 7-38. Typical amplifier circuit using the MC1306P power amplifier.

operational amplifiers, and some of the same principles apply.

Figure 7-38 shows a typical circuit connection. The left amplifier, A, is the preamplifier, and the right amplifier, B, is the. power amplifier. Using this circuit, the voltage gain of the preamplifier is given by the equation:

$$A_{VA} = R_f/R_s$$

And the gain of the amplifier circuit is given by:

$$A_{VB} = 10\ K/R_p$$

The 10 kilohm resistor connected as the feedback resistor for the power amplifier is provided in the integrated circuit. The recommended values of R_p are from 500 ohms to 3.3 kilohms, giving a gain

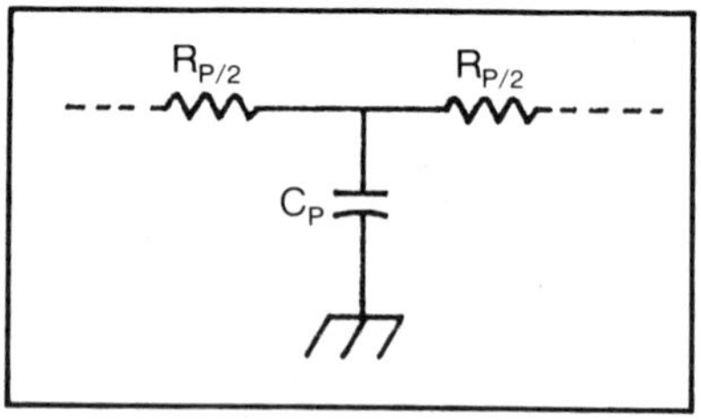

Fig. 7-39. Dividing R_p and adding a compensation capacitor.

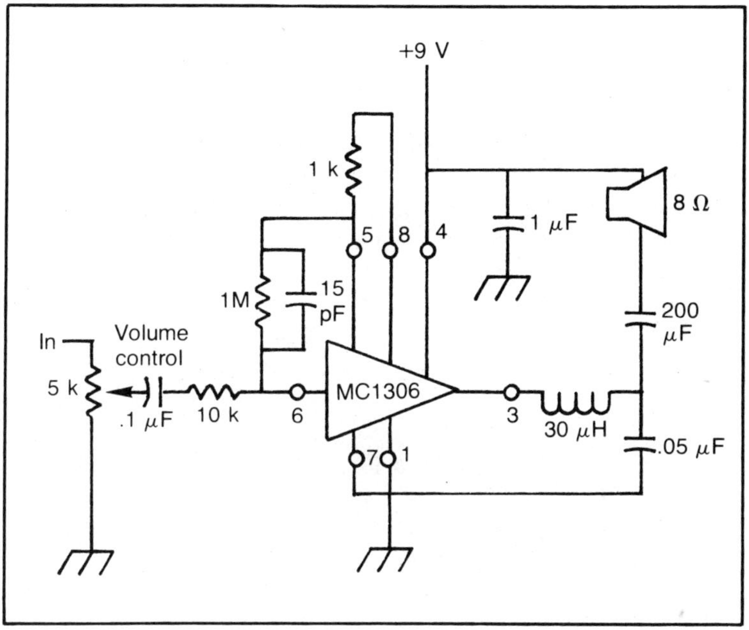

Fig. 7-40. Typical audio amplifier circuit.

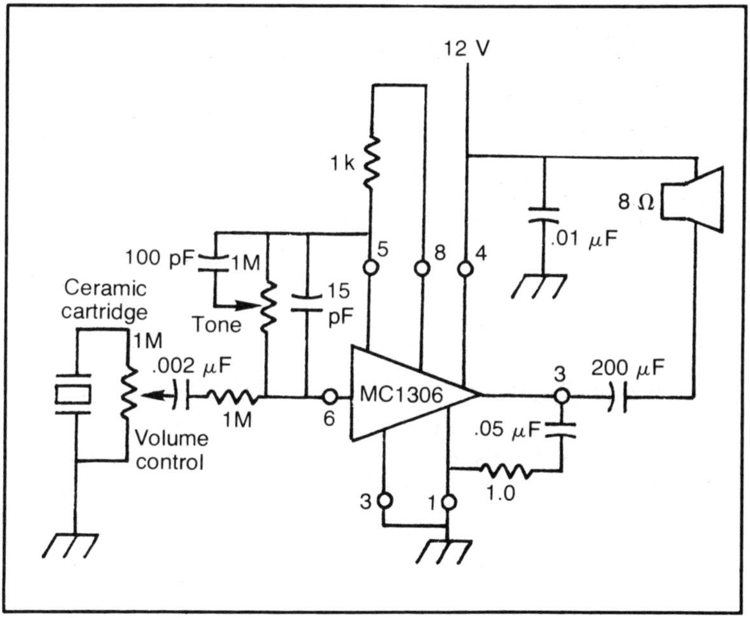

Fig. 7-41. Amplifier circuit for phono cartridge.

of from 20 to 3. The low resistance end is limited primarily by low level distortion and the high resistance end is limited by the voltage drive capabilities of the preamplifier.

Both of the above gain equations are only approximations, because of the low open-loop gain. Although they are not exact, they are close enough for the design of the circuit.

The overall gain of the circuit is the gain of the preamplifier times the gain of the power amplifier and is given by the following equation:

$$A_{Vtotal} = A_{VA} \times A_{VB} \quad \frac{R_f \times 10\ K}{R_s \times R_p}$$

The input impedance of the preamplifier is essentially R_s. The input impedance of the power amplifier is approximately R_p.

The low frequency response is controlled by the cumulative effect of the series coupling capacitors C1, C2, and C3. The high frequency response can be determined by the feedback capacitor C_f, and the -3 dB point occurs when:

$$2\pi \times F \times C_f = R_f$$

Additional high frequency rolloff and the noise reduction can be achieved by placing a capacitor from the center point of R_p to ground, as shown in Fig. 7-39.

A 470 pF capacitor may be required from the power amplifier end of R_p to ground to prevent high frequency oscillations.

Figure 7-40 shows a typical audio amplifier application such as that used in a radio. This circuit has a preamplifier gain of 100 and a power amplifier gain of 10. This gives an overall voltage gain of 1000. The bandwidth is 10 kHz.

The application in Fig. 7-41 shows an amplifier for a ceramic phono cartridge. This circuit has a preamplifier gain of unity, and a power amplifier gain of 10 for an overall gain of 10. A gain of 10 is sufficient because of the high level output signal from the ceramic cartridge. The input impedance to the preamplifier is 1 megohm. A tone control adjustment is provided within the preamplifier feedback loop.

8

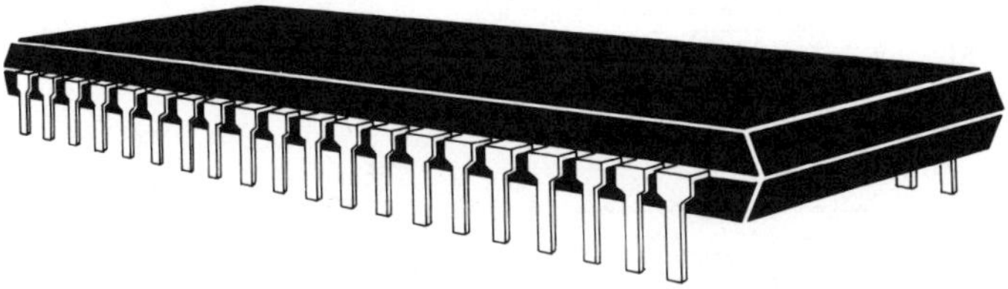

Projects

The projects in this chapter are designed to acquaint the reader with using integrated circuits in actual, working circuits. They are simple projects with relatively few components.

You are guided through the assembly of the projects by a wire list and a parts list. The theory of operation is also described. The expected results and any operational variations are discussed. This chapter gives you some hands on experience using integrated circuits.

Before starting the projects, there are some general topics which must be covered. The first is the powering of the projects. Most of the projects can be powered from a 5 volt power supply. Realizing that a 5 volt power supply may no be available, there are two alternative methods of powering the projects.

The first method is to use a 6 volt battery, and some diodes, as shown in Fig. 8-1. Two diodes are used in the forward direction so that the diode voltage drop subtracts from the battery voltage. This is enough to drop the 6 volts to a value within the limits of TTL integrated circuits. Using two general purpose silicon diodes, such as the IN4000 series, drops the voltage to within the limits. If a meter is available, it is advisable to measure the voltage under load, as shown in Fig. 8-2. A 1 kilohm resistor will simulate the operational load. If the voltage is high, add more diodes until it is about 5 volts.

A second, and recommended, method is to use a 9 volt tran-

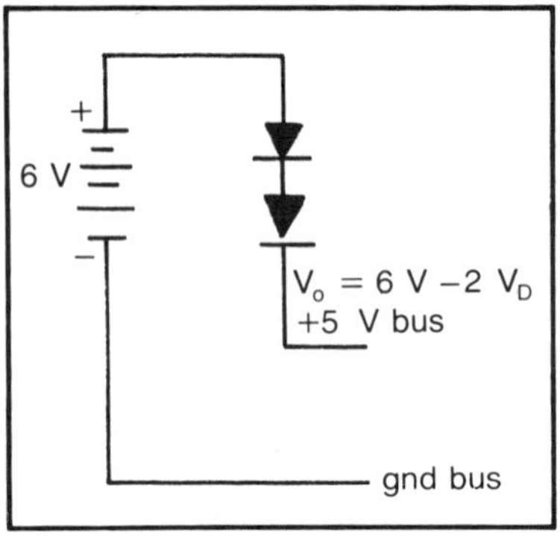

Fig. 8-1. Using the forward voltage drop of a diode to drop 6 volts to about 5 volts.

sistor radio battery, and a 5 volt three-terminal regulator, to develop the 5 volts. This is the first project, and is covered in detail later.

A method of assembling the circuits must be used. I do not expect you to construct the projects by soldering them together on a circuit board of some type. It is best to assemble them on some type of test board. Using a test board, the connections can be quickly removed and connected. There are several types of test, or experimenters boards that can be purchased or made. The most convenient type is the prototype plug in board, as shown in Fig. 8-3. Several manufacturers make them, and they are available in different price ranges. Figure 8-4 gives the general layout of such a board. There are two parts to it, the plug in block and the bus strip. These may be separate pieces, or part of the same unit. The bus strip has two rows of terminals going the length of the strip. Each row of terminals is bussed together, and the two rows are electrically separate. These two strips of terminals are for the ground and voltage busses.

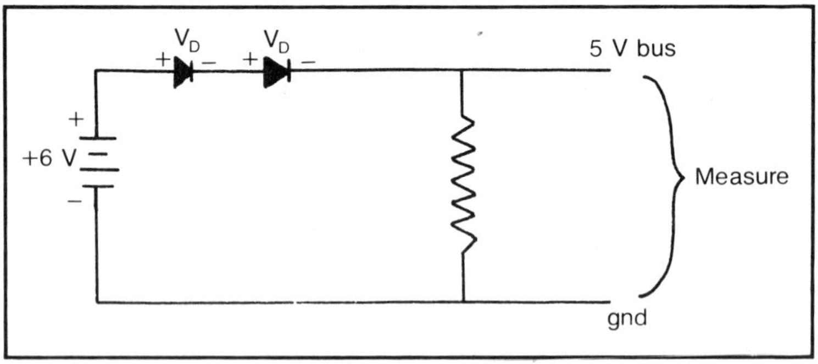

Fig. 8-2. Measuring the voltage from the diodes.

Fig. 8-3. Photograph of a plug-in test board.

The plug in block consists of several rows of 4 or 5 terminals on each side of a center gap. The rows of 4 or 5 terminals are connected together, and electrically separated from the other rows of terminals. These rows of connected terminals go from the center gap to the sides of the block.

The terminals for the plug in board are small pin sockets located inside the holes in the block. These are sized for a 22 to 30 gauge wire. Just strip off a ¼ inch of insulation and poke the bare end into the hole. Pull slightly on the wire to make sure that the terminal is gripping the wire securely. The integrated circuits are plugged in across the center gap. All the pins on one side of the IC are on one side of the gap and all the pins on the other side of the IC are on the other side of the gap. The pins are isolated from each other. Short pieces of wire, stripped at both ends, are used to provide the interconnection. Components with leads small enough to fit in the

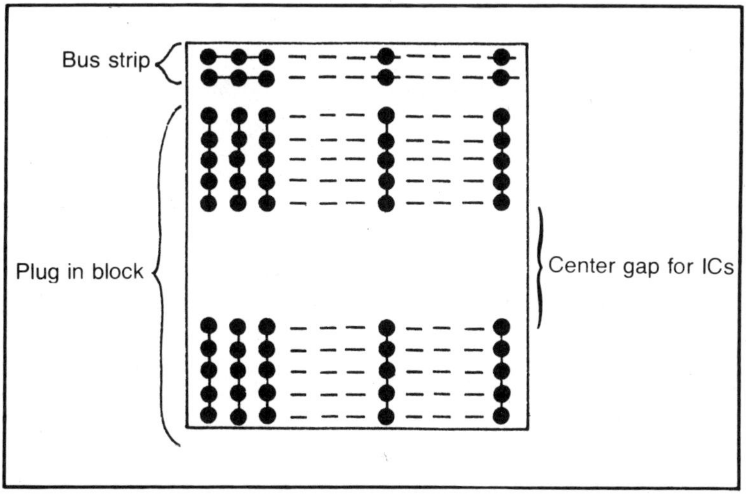

Fig. 8-4. Terminal layout of the plug-in test board showing which terminals are bussed together.

terminals, can be plugged into the terminals. Components with larger diameter leads require the use of clip leads, or require a smaller diameter wire be soldered to it before plugging it into the terminal.

When placing components on the test board, remember that the rows of terminals on each side of the center gap are connected together. If two wires or leads are plugged into the same row of terminals, they are electrically connected together. Only one wire can be inserted into each terminal.

Another type of test board is shown in Fig. 8-5. It consists of a 16 pin integrated circuit socket, and three rows of double terminal strips, all mounted on a common mounting board. The leads of the components, and the interconnecting wire is attached to the screw terminals of the terminal strips. A 16-pin IC socket is used because an 8-pin, 14-pin and 16-pin IC will all fit into this socket.

Another type of test board is made from spring clips. The leads of components and the interconnecting wiring are held by the clips.

There is always the option of soldering the components to the IC socket pins, in a three dimensional type of arrangement. Interconnecting wiring is used where required. Soldering is difficult and time consuming, and makes it difficult to change components and wiring. Repeated soldering can damage the components and make the leads brittle. However, many electronic hobbyists use this method for experimental circuits.

A better method is to use clip leads to connect the components together. Small spring loaded clips, or small alligator clips can be

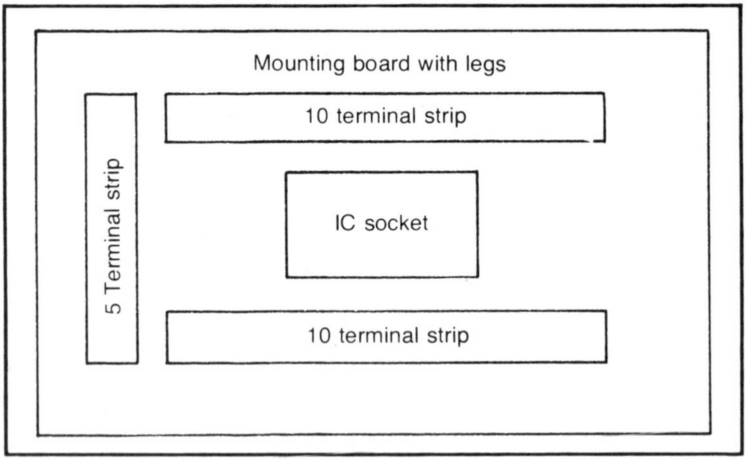

Fig. 8-5. A test board made from terminal strips.

used to make up several short clip leads, between 4 to 6 inches long. Solder some short leads to the IC socket to facilitate connecting the clip leads. Be careful when interconnecting the circuit. It is easy to create shorts in the circuit. It is also easy for the clip leads to come off, creating an open.

I recommend the plug in board mentioned first. All things considered, this will be the cheapest in the long run. It requires no soldering and no extra clip leads. It is convenient to use, and easy to change. It is the method that I use.

A multimeter is very handy for working with electronic circuits. In fact, it is a must for the serious hobbyist. A multimeter can measure ac voltage, dc voltage, current, and resistance. A good multimeter can be purchased for less than $15, and is well worth the investment, although it is not required to complete the projects in this chapter.

For many of the projects, a light emitting diode (LED) is used to indicate the status of the output. The LED is a handy device, in that it will indicate current flow through the diode. When there is sufficient current flow, the LED will glow. Otherwise it will be dark. To obtain maximum brightness a standard LED requires 16 mA current, whereas a typical bulb requires at least 60 mA. Typically, 5 mA will illuminate the LED sufficiently to see that it is on.

An LED has a positive and a negative terminal. If the positive terminal is to the positive side of the current source, and the negative terminal to the negative side, the LED will illuminate. If the terminals are reversed, the LED will act like an ordinary diode, conducting current. Figure 8-6 shows a simple circuit for determining the polarity of the LED. If the LED does not glow, reverse the connections. When the terminals are identified, mark them so

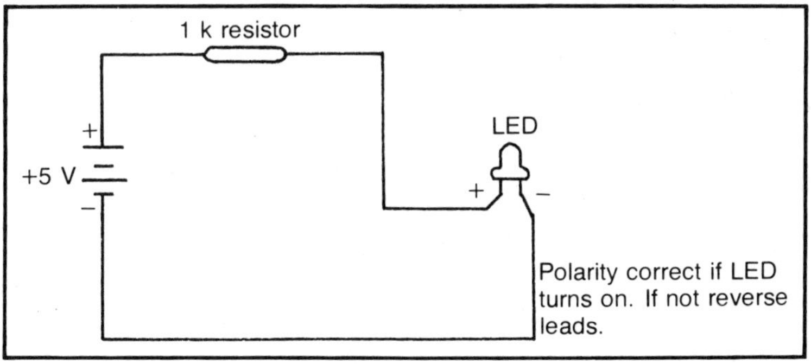

Fig. 8-6. Checking the polarity of the LED to find the positive lead.

that they can be connected into the circuit correctly. Whenever the LED is connected to a current source, there must be a current limiting resistor to limit the current to between 5 and 16 milliamps. This will prevent destroying the LED.

LEDs come in many sizes and colors. The colors are typically red, green, yellow, and amber. It does not matter which color is selected because their characteristics are all about the same. The size determines the current required, so it is best to use something about the T-2 size. This is about 0.2 inches in diameter, and about twice as high. It will emit sufficient light with 5 milliamps to be noticeable.

The parts required for the projects, including the test board, are available from an electronics hobbyist store or from a mail order electronics parts supplier. Be careful to get only good components, and not second rate or defective components. Some surplus electronic stores have good components at a discount price, while others have only junk.

The part numbers for the ICs are given in the project parts list. Some part numbers are peculiar to certain manufacturers, while other manufacturers have equivalent parts with a slightly different part number. This applies particularly to the characters at the start of the part number, such as MC, LM, etc. Most of the time, the numbers for equivalent parts are the same, but there are cases where the numbers of two parts are the same and the letters are different, but they are totally different parts. Be careful when substituting part numbers and make sure you get the equivalent part.

Several reputable electronics mail order companies advertise in the electronics magazines. Consult one of these magazines for addresses and send for catalogs. Read the advertisement or the catalog carefully to determine if the parts are new, first class parts or if they are factory seconds or rejects.

Read through each project carefully, and be sure you understand what is to be accomplished. Understand how the circuit is set up and what the expected results are. Refer to the previous chapters to review the operation of the integrated circuit, and understand how it works in the circuit being assembled. Then assemble the circuit, but do not connect the power. Double check the wiring to make sure that it is correct and that there are no shorts. Then connect the power.

In the parts list for the projects, the test board is not defined. The choice of a test board is up to you.

	C1	R1	Reg Pin	Bat Clip
5 V Bus	+	Top	Out	
Gnd	–	Low	Gnd	–
			In	+

Table 8-1. Wire List for Project 1

PROJECT 1, THE 5 VOLT REGULATOR

This project uses the LM340T-5 three-terminal regulator to regulate the 9 volts from the battery to 5 volts to power most of the following projects.

The parts required are:

9 volt transistor radio battery
Battery clip for the battery
LM340T-5 three-terminal regulator, 5 volt
0.01 μF ceramic disk capacitor
470 ohm resistor, ¼ watt

Figure 8-7 is the schematic for the regulator, and Table 8-1 is the wiring list. In a wire list, each line represents a function and each column represents a component. All the connections in one line are common. For instance, the first line of the wire list says that the 5 volt bus, which is the output, is connected to the out pin of the

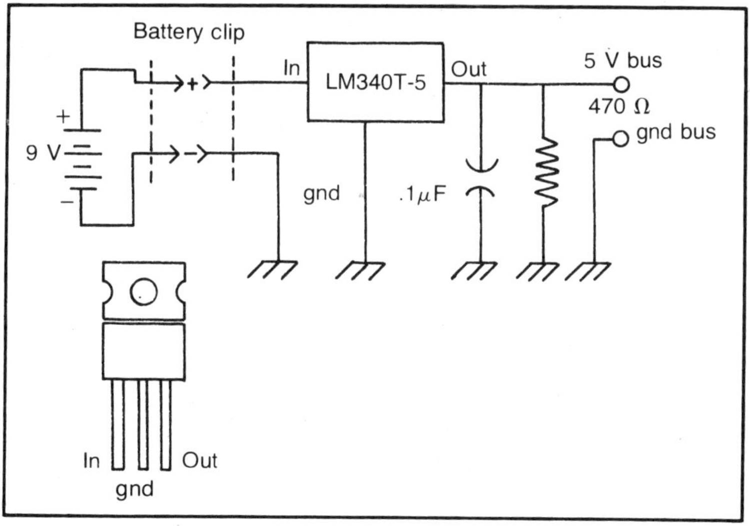

Fig. 8-7. LM340T-5 regulator used to regulate a 9 volt battery to 5 volts.

regulator, the top of R1, and the + side of the capacitor. Resistors do not have a top or bottom, but top and low are used to designate the two ends of the resistor. A ceramic disk capacitor is not polaried, so the + and – signs indicate the two leads.

A battery clip can be made from an old discarded 9 volt transistor radio battery. Carefully remove the top with the snap terminals. Then solder some wire to the terminals to connect to the circuit. Realize that the terminals on the battery are reversed from those on the clip. The positive terminal on the clip was the negative terminal on the battery.

Assemble the circuit on one corner of the test board because it will remain in use for most of the projects. Do not connect the battery clip to the battery until the wiring has been double checked. If there is a voltmeter available, connect it across the 470 ohm resistor. If a voltmeter is not available, connect the positive terminal of a LED to the 5 volt bus, and connect the negative terminal of the LED to ground through a 1 kilohm resistor. Then connect the battery clip to the battery. The voltmeter should read between 4.8 and 5.2 volts, or the LED should glow. If this does not happen, disconnect the battery and check the wiring again. Also check the 9 volt battery.

The 470 ohm resistor provides a load for the regulator when checking out the circuit. When the circuit is powering a circuit, remove this resistor. The 5 volt bus and the ground bus are as marked on the schematic, and should be connected to the bus points on the test board.

PROJECT 2, NAND GATE

This project demonstrates the operation of a single NAND gate using the LED as an indicator. The parts required are:

Project 1, assembled
7400 Integrated Circuit
LED
2 1K resistors, ¼ watt

Figure 8-8 is the schematic for the NAND gate project, and the wire list is given in Table 8-2. Note that only the first NAND gate in the IC is used, but any of the four gates can be used by changing the input and output pin numbers. The LED is used as an indicator and R1 is the current limiting resistor for the LED. Since an LED is on when there is current flow, it will be on when the output of the

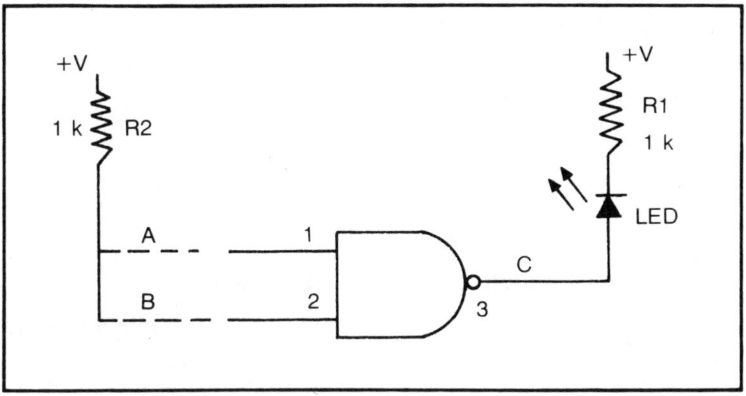

Fig. 8-8. The schematic for Project 2.

NAND gate is low. Resistor R2 is used to generate a high level for the inputs. To make an input high, connect it to the lower end of R2. To make an input low, connect it to ground. This creates a high and low logic level well within the values required for the IC operation. Start with the inputs connected to the resistor.

Connect the circuit as shown, leaving the battery clip disconnected. double check the wiring and make sure there are no shorts, then connect the battery clip. The LED should be on. If it is not, disconnect the battery and check the circuit again.

Wires A and B on the schematic are the gate inputs. They are initially set high by the resistor, giving two high inputs to the NAND gate. The result is a low from the gate, which illuminates the LED. This is the first line of the truth table for this circuit, which is given in Table 8-3. Step through the lines of the truth table, one line at a time. This is accomplished by connecting inputs A and B according to the appropriate columns, and observe the LED. Notice that the only time the LED is on is when both inputs are high. This is the function of a NAND gate.

Table 8-2. Wire List for Project 2

	R1	R2	LED	IC Pin
5 V Bus		Top	+	14
Gnd				7
	Top		–	
	Low			3
		Low		1,2

A	B	C	LED
1	1	0	On
1	0	1	Off
0	1	1	Off
0	1	1	Off

Table 8-3. Truth Table for Project 2

PROJECT 3, SERIES NAND GATES

This project uses two NAND gates in series to illustrate a slightly more complex gate. The parts required are:

Project 1, assembled
7400 integrated circuit
LED
2 1K resistors, ¼ watt

This project may be assembled by adding only a few wires to the first project. Figure 8-9 shows the schematic for this project. The wire list is given in Table 8-4.

This logic circuit has three inputs: A, B, and C. The connections given in the wire list should result in the LED being on.

Assemble the circuit on the test board, and check the wiring for shorts and opens. When the wiring is correct, connect the battery. The LED should be on.

Table 8-5 gives the truth table for this project. The circuit, as assembled, will be the first line of the truth table. Step through the

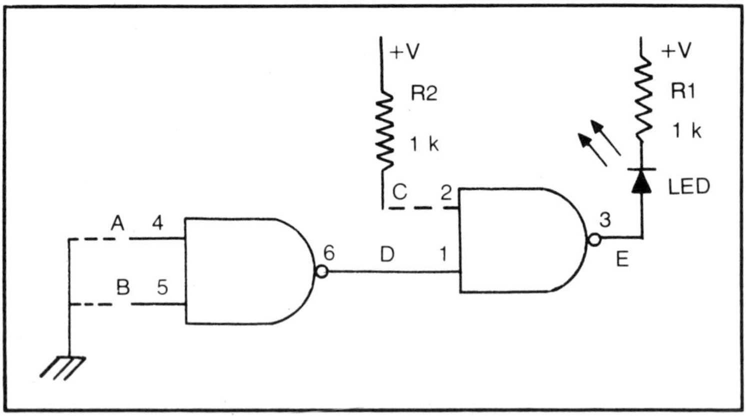

Fig. 8-9. The schematic for Project 3.

Table 8-4. Wire List for Project 3

	R1	R2	LED	IC Pin
5 V Bus		Top	+	14
Gnd				7,4,5
	Top		–	
	Low			3
		Low		2
				6,1

truth table, one line at a time, connecting the inputs as shown in columns A, B, and C. Column D gives the intermediate results from the first gate. Column E gives the final output, which when low, illuminates the LED.

For a 1 input, connect the proper input to the lower end of R2. For a 0 input, connect it to ground. Step through the truth table and observe the LED. Note that input C must be high for the LED to be on. Signal D must also be high for the LED to be on. This is not surprising because it is the basic operation of the NAND gate. From the truth table the logic equation, can be written for both the high output state and low output stage by determining all the conditions required to achieve that condition.

$$E = C + (A \bullet B)$$
$$E = C \bullet (\overline{A} + \overline{B})$$

Table 8-5. Truth Table for Project 3

A	B	C	D	E	LED
0	0	1	1	0	ON
0	1	1	1	0	ON
1	0	1	1	0	ON
1	1	1	0	1	OFF
1	1	0	0	1	Off
0	1	0	1	1	Off
1	0	0	1	1	Off
1	1	0	0	1	Off

The first equation states that for output E to the high, input E must be low or both inputs A and B high. The second equation states that for output E to be low, input C must be high and either input A or B low. Check these equations against the truth table. They fit all conditions. Since these are the only two conditions of signal E, the equations must be the complement of each other. That is, the second equation covers all the conditions not covered by the first. This can be proven by boolean algebra, which is the mathematics which deals with logic equations.

PROJECT 4, WIRED AND GATE

This project uses two NAND gates to generate a wired AND gate. Recall from Chapter 6 parallel outputs connected to a common point perform an AND function. The two NAND gates provide the inputs to the wired AND gate, while another NAND gate is used as an inverter to simulate the driven circuit and to drive the LED. The parts required are:

Project 1, assembled
7400 integrated circuit
2 1 K resistors, ¼ watt
LED

Figure 8-10 is the schematic for this project, and Table 8-6 is the wire list. Signal E is the wired AND gate that feeds the inverter. The inverter drives the LED. For this circuit, if the LED is on, it indicates that the AND gate is high.

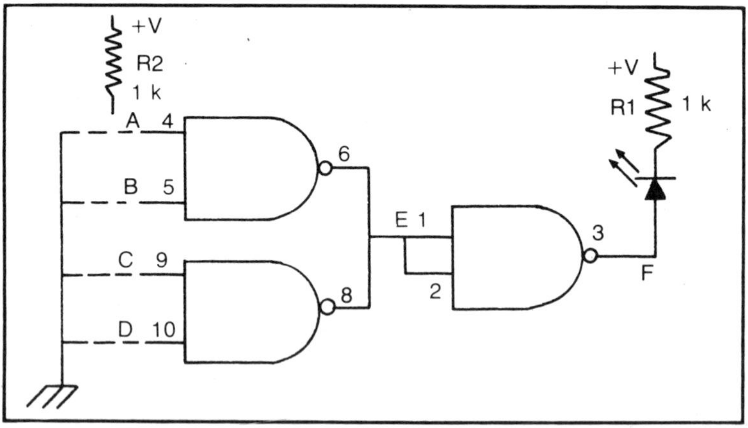

Fig. 8-10. The schematic for Project 4.

Table 8-6. Wire List for Project 4

	R1	R2	LED	IC pin
5 V Bus		Top	+	14
Gnd Bus				7,4,5,9,10
	Top		–	
	Low			3
				1,2,6,8

Assemble the circuit according to the wire list, and check the wiring. After everything is checked out, connect the battery. The LED should be on. If it is not, disconnect the battery and check the wiring again.

The truth table for this circuit is given in Table 8-7. Note that there are two columns which have not been included. These are the columns for the outputs of the two parallel NAND gates, pins 6 and 8. You can add these if you want to. Realize that this type of connection, as discussed in Chapter 6, results in an interaction

Table 8-7. Truth Table for Project 4

A	B	C	D	E	F	LED
0	0	0	0	1	0	On
1	0	0	0	1	0	On
0	1	0	0	1	0	On
1	1	0	0	0	1	Off
0	0	1	0	1	0	On
0	0	0	1	1	0	On
1	0	0	1	1	0	On
0	1	0	1	1	0	On
1	1	0	1	0	1	Off
0	0	1	1	0	1	Off
1	0	1	1	0	1	Off
0	1	1	1	0	1	Off
1	1	1	1	0	1	Off

between the two outputs. If one of the outputs are low, it pulls the remaining outputs low, resulting in a low at the AND gate.

Step through the truth table, one line at a time by connecting the inputs as shown under columns A, B, C, and D. Observe the LED, and make sure the indication is as specified in Table 8-7. Remember the function of the NAND gate. If both inputs are high, the output is low. Otherwise the output is high. To achieve a high at both connections of the wired AND gate requires that both NAND gates satisfy the condition for high output. So, when both inputs to one or both of the NAND gates are high, the AND gate (signal E) will be low, resulting in the LED being off. When one or more of the inputs to both NAND gates are low, the AND gate will be high, resulting in the LED being on.

PROJECT 5, NOR GATE

This project is to demonstrate the operation of a two input NOR gate. The parts required are:

Project 1, assembled
7402 integrated circuit
LED
2 1 K resistors, ¼ watt

The schematic for this project is given in Fig. 8-11, and the associated wire list is in Table 8-8. This circuit is similar to that of project 2, except that a different IC is used. The 7402 is a 2-input NOR gate containing 4 independent gates.

Assemble the circuit on the test board using the wire list. Carefully check the wiring before connecting the battery. Then connect the battery clip to the battery. The LED should be on.

The truth table is given in Table 8-9. The circuit, as assem-

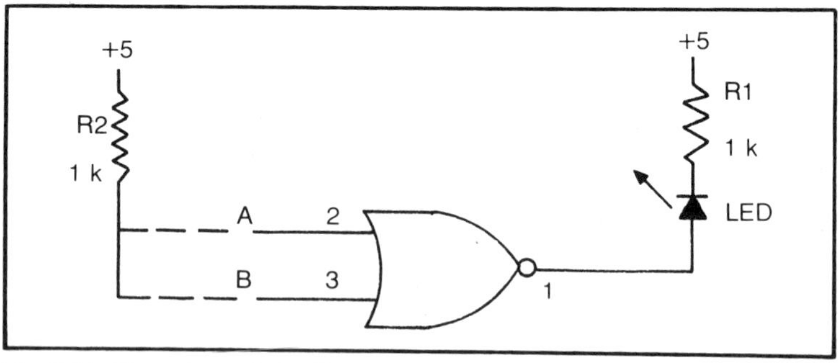

Fig. 8-11. The schematic for Project 5.

Table 8-8. Wire List for Project 5

	R1	R2	LED	IC Pin
5 V Bus	Top	Top		14
Ground				7
	Low		+	
			−	1
		Low		2,3

bled, is the first line of the truth table; therefore, the LED will be on when you apply power.

Step through the lines of the truth table, one line at a time, by connecting the inputs as shown in columns A and B. Look for the proper LED response. Notice that the only time the LED is off, indicating a high output, is when both inputs are low.

PROJECT 6, SERIES NOR GATES

This project uses two of the NOR gates of the 7402 IC, connected in series, to demonstrate the series operation of NOR gates. The parts required are:

Project 1, assembled
7402 integrated circuit
LED
2 1 K resistors, ¼ watt

The schematic for this project is shown in Fig. 8-12, and the associated wire list is given in Table 8-10. Since there are only a few additional wires required from the previous project, this project can be wired without dismantling project 5.

Wire the circuit on the test board using the wire list. When the

A	B	C	LED
1	1	0	On
0	0	1	Off
0	1	0	On
1	0	0	On

Table 8-9. Truth Table for Project 5

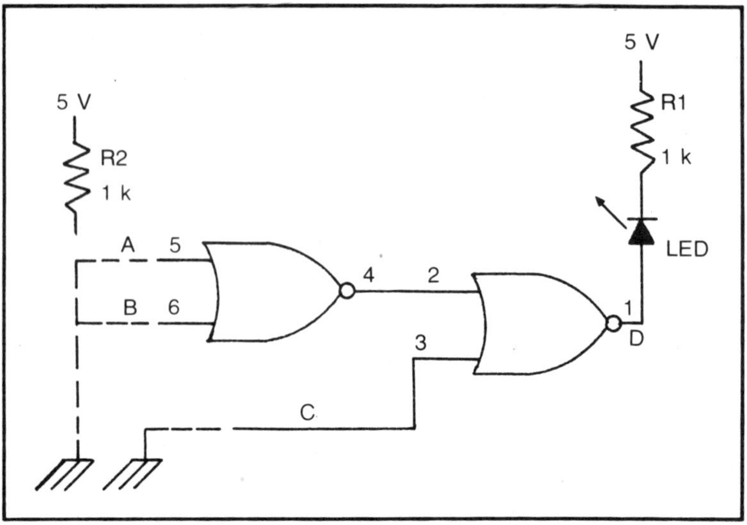

Fig. 8-12. The schematic for Project 6.

wiring has been checked, connect the battery. The LED should come on. If it does not, recheck the wiring.

The truth table is given in Table 8-11, with the circuit reflecting the first line. Step through the truth table, one line at a time by changing the inputs as specified in columns A, B, and C. Note that anytime input C is high, the LED is on. Also note that anytime both A and B are low, the LED is on. From these observations, the equations for the circuit can be written.

$$D = \overline{C} \cdot (A + B)$$
$$D = C + (\overline{A \cdot B})$$

These equations say the same thing as the above statements, and they are complementary.

	R1	R2	LED	IC Pin
5 V Bus	Top	Top		14
Gnd				7,3,5,6
	Low		+	
			−	1
				2,3

Table 8-10. Wire List for Project 6

Table 8-11. Truth Table for Project 6

A	B	C	D	LED
0	0	0	0	On
1	0	0	1	Off
0	1	0	1	Off
1	1	0	1	Off
0	0	1	0	On
1	0	1	0	On
0	1	1	0	On
1	1	1	0	On

PROJECT 7, AND GATE

This project is to demonstrate the operation of a single, two input AND gate. The parts required are:

Project 1, assembled
7408 integrated circuit
LED
2 1 K resistors, ¼ watt

The schematic for the circuit is given in Fig. 8-13, and the associated wire list is given in Table 8-12. Assemble the circuit,

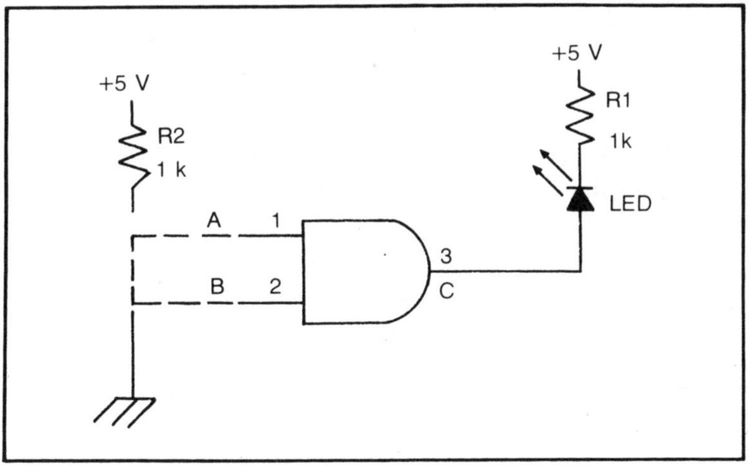

Fig. 8-13. The schematic for Project 7.

	R1	R2	LED	IC Pin
5 V Bus	Top	Top		14
Gnd				7,1,2
	Low		+	
			−	3

Table 8-12. Wire List for Project 7

check the wiring, and connect the battery. The LED should be on. If it is not, disconnect the battery and correct the wiring.

The truth table for the AND gate is given in Table 8-13. Compare this truth table with Table 8-3, which is the truth table for a NAND gate. The only difference is that the output, column C, and the LED indications are inverted. This is to be expected, since a NAND gate is the equivalent of an AND gate with an inverter.

Step through the truth table by changing the inputs as defined in columns A and B, and observe the LED indications. Both inputs A and B must be high for the output to be high and the LED off. This is as expected by the definition of an AND gate.

PROJECT 8, SERIES AND GATE

This project connects all four of the AND gates in a series and parallel circuit in a more complex logic circuit. The parts required are:

Project 1, assembled
7408 integrated circuit
LED
2 1 K resistors, ¼ watt

The schematic for this circuit is given in Fig. 8-14, and the associated wire list is given in Table 8-14. The two left AND gates

Table 8-13. Truth Table for Project 7

A	B	C	LED
0	0	0	On
0	1	0	On
1	0	0	On
1	1	1	Off

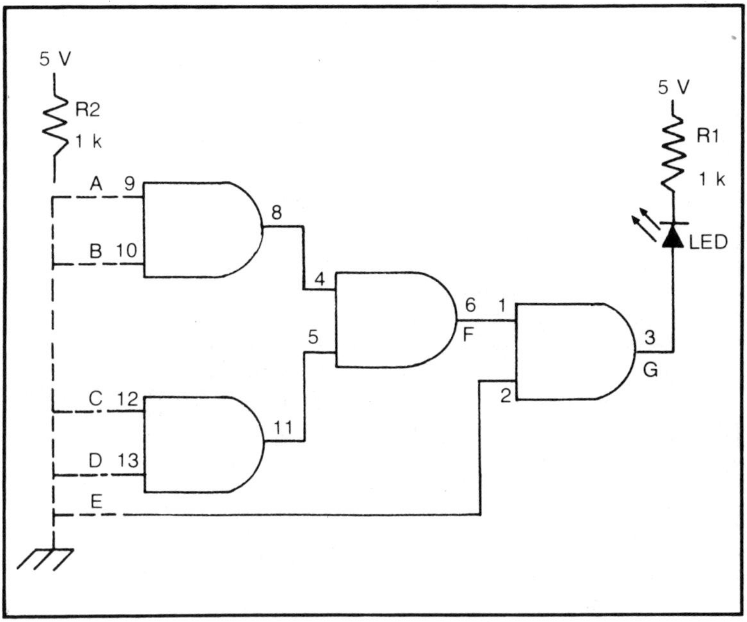

Fig. 8-14. The schematic for Project 8.

are connected in parallel, into the center AND gate, which is connected to the remaining AND gate along with another input. The circuit has five inputs: A, B, C, D, and E.

Assemble the circuit on your test board according to the wire list. Double check the wiring, and then connect the battery. The LED should be on.

Table 8-14. Wire List for Project 8

	R1	R2	LED	IC Pin
5 V Bus	Top	Top		14
Gnd				7,9,10,12,13,2
	Low		+	
			–	3
				8,4
				11,5
				6,1

A	B	C	D	E	F	G	LED
0	0	0	0	0	0	0	On
1	0	0	0	0	0	0	On
0	1	0	0	0	0	0	On
1	1	0	0	0	0	0	On
0	0	1	0	0	0	0	On
1	0	1	0	0	0	0	On
0	1	1	0	0	0	0	On
1	1	1	0	0	0	0	On
0	0	0	1	0	0	0	On
1	0	0	1	0	0	0	On
0	1	0	1	0	0	0	On
1	1	0	1	0	0	0	On
0	0	1	1	0	0	0	On
1	0	1	1	0	0	0	On
0	1	1	1	0	0	0	On
1	1	1	1	0	1	0	On
0	0	0	0	1	0	0	On
1	0	1	0	1	0	0	On
1	1	1	1	1	1	1	Off

Table 8-15. Truth Table for Project 8

The truth table for this circuit is given in Table 8-15. Notice that there are 13 lines missing from this truth table. These fall between the three lines at the bottom of the table. They are left out so that you can determine them. Columns A, B, C, and D will be repeats of the above lines with columns F and G being all zeros except for the last line, where all the inputs are high.

Sequence through the truth table as given, one line at a time, and notice the LED response. The LED should be on for all conditions when all the inputs are high. For the LED to be off, signal G must be high, which means that E and F must be high. For F to be high requires that A, B, C, and D be high. All this circuitry has made a 5 input AND gate where all the inputs must be high for the output to be high. The logic equation for this circuit is as follows:

$$G = A \cdot B \cdot C \cdot D \cdot E$$
$$G = A + B + C + D + E$$

As an exercise replace the 7408 AND gate with the 7400 NAND gate, and determine the results for the same inputs. Columns F, G, and the LED indications will be inverted.

PROJECT 9, LATCH CIRCUIT

The latch circuit is a logic circuit which, when set to one condition,

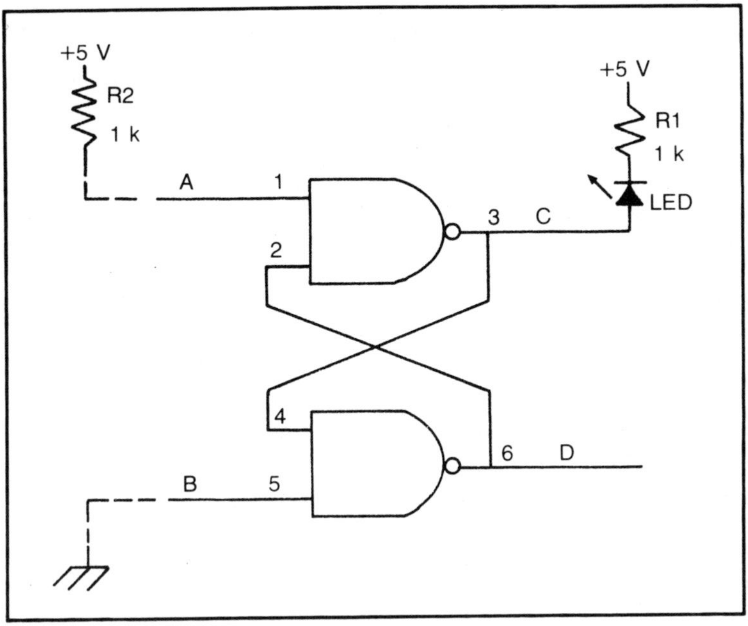

Fig. 8-15. Schematic for Project 9.

will remain in that condition until it is set to the other condition. As discussed in Chapter 6, the latch has two stable states, the set and reset states. In this project, the latch circuit is assembled and sequenced through the various conditions. The parts required are:

Project 1, assembled
7400 integrated circuit
LED
2 1 K resistors, ¼ watt

Table 8-16. Wire List for Project 9

	R1	R2	LED	IC Pin
5 V Bus	Top	Top		14
Gnd				7,5
	Low		+	
			–	3,4
		Low		1
				2,6

The circuit is shown in Fig. 8-15. The wire list is given in Table 8-16. The circuit is essentially two NAND gates connected back to back, with an external input for each gate. These inputs are the set and reset terminals. Making one of the inputs active (low) will place the circuit in that condition. When that input is returned high, the circuit will stay low until the other input is made active. Then the circuit will switch to the other condition. The upper input (A) is the set input, and the upper output (C) is the set output, which is high when the circuit is in the set condition.

Assemble the circuit and check for correct wiring. When you connect the battery, the LED should be on. If it is not make sure B is grounded, and check the wiring.

Table 8-17 gives the truth table for the latch circuit. For the original condition, as per the wire list, the reset input is active, resetting the latch. Therefore the upper output is low, illuminating the LED. The next step in the truth table returns both inputs high, and showing that the circuit does not change state. Since this is a sequential truth table, the steps must be done in sequence for the indications to be correct; therefore, the input leads must be moved while the circuit is powered up. If the circuit switches while moving the leads, return to the original state and start over. A little care will allow the circuit leads to be changed.

Step through the lines on the truth table and observe the results. Set and reset the circuit by using lines 1 and 3 of the truth table. Then set the circuit and let it remain in that condition for a little while to demonstrate that the circuit will maintain the condition. Play with the circuit until the operation is thoroughly understood.

When this latch is fully understood, replace the 7400 NAND gate with the 7402 NOR gate and determine the operation. Before applying power, determine the active states of the inputs and the responses. Remember that the active state for the NAND gate latch was low and was determined by the fact that any low input for a

A	B	C	D	LED
1	0	0	1	On
1	1	0	1	On
0	1	1	0	Off
1	1	1	0	Off

Table 8-17. Truth Table for Project 9

NAND gate results in a high output. For the NOR gate, any high input results in a low output. This makes the inputs active high and the outputs active low. Proceed from here by drawing the circuit and deriving the truth table. Then set up the circuit and sequence through the truth table. If everything was done correctly, the results will match the truth table.

Realize that the active state of the output is determined by which output is labeled the set output. For Fig. 8-15, the upper output was called the set output, making the output active high. But, if the lower output (D) was called the set output, the signal will be active low.

Next, use the 7408 AND gate in the latch circuit. Determine the active states and the truth table. This exercise illustrates the use of different gates in the same circuit and gives three latch circuits with three different input and output conditions. It is also a start at logic circuit analysis using the truth table.

PROJECT 10, COMPLEX LOGIC CIRCUIT

This project demonstrates the use of all three of the digital ICs in one circuit. The parts required are.

Project 1, assembled
7400 integrated circuit
7402 integrated circuit
7408 integrated circuit
LED
2 1 K resistors, ¼ watt

The schematic for this project is shown in Fig. 8-16, and the associated wire list is given in Table 8-18. The circuit consists of a latch circuit with a NOR gate on the reset line. A combination of AND and NOR gates are on the set side of the latch. The object of the circuit is to set and reset the latch so that the output may be a high when the set combination is achieved

Assemble the circuit according to the wire list. Since this is the most involved circuit yet assembled be careful to assure that the wiring is correct and that there are no shorts. When the circuit has been checked, connect the battery. The LED should come on. Referring to the circuit diagram, the initial state of the circuit is with the reset input to the latch (J) active. This condition results in output K being low, and the LED on. In this manner, the circuit is powered up in a known condition.

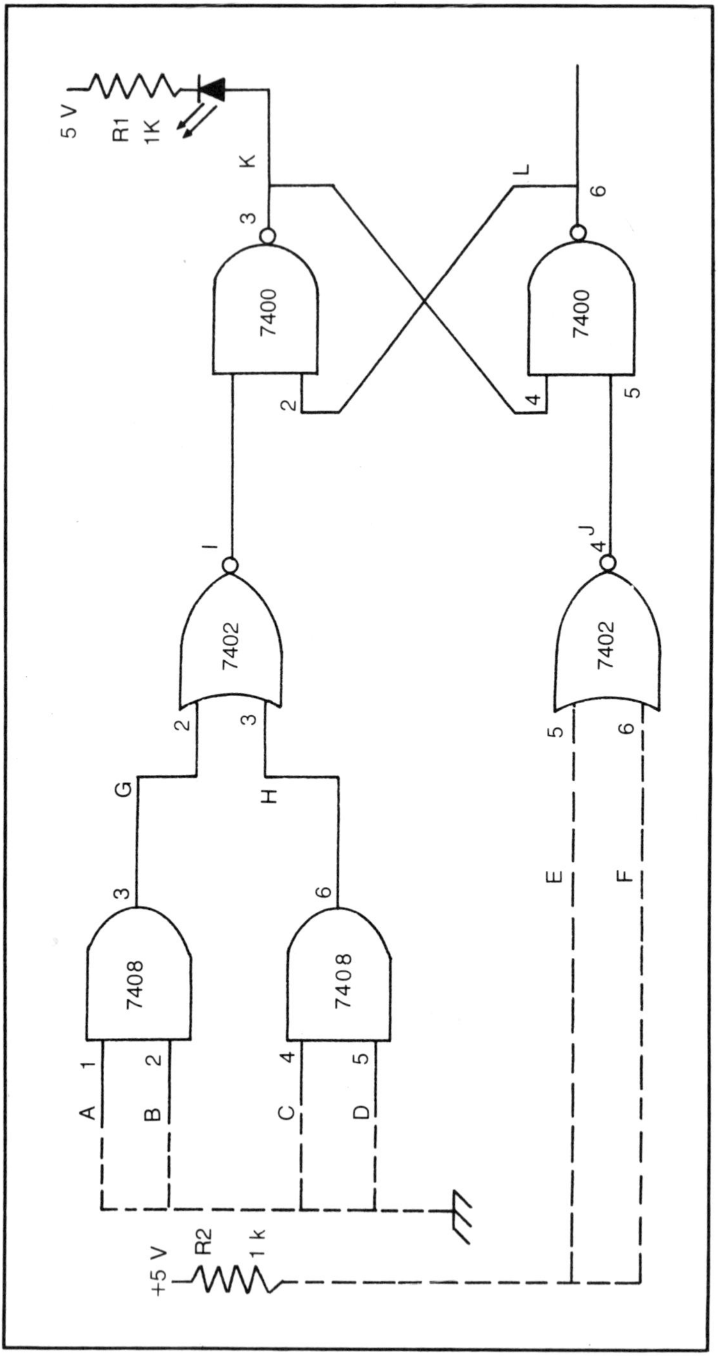

Fig. 8-16. Schematic for Project 10.

Table 8-18. Wire List for Project 10

	R1	R2	LED	7400	7402	7408
5 V Bus	Top	Top		14	14	14
Gnd				7	7	1,2,4,5,7
	Low		+			
			–	3,4		
				2,6		
				1	1	
				5	4	
					2	3
					3	6
		Low			5,6	

Table 8-19 is the truth table for this circuit. The inputs, connected per the wire list, are set up for the first line. The second, third, and fourth lines are signal inputs which do not affect the latch, but the fifth line sets the latch, and line 6 resets the latch. Sequence through the truth table, one line at a time, and check for the proper

Table 8-19. Truth Table for Project 10

A	B	C	D	E	F	G	H	I	J	K	L	LED
0	0	0	0	1	1	0	0	1	0	0	1	On
0	0	0	0	0	0	0	0	1	1	0	1	On
1	0	0	0	0	0	0	0	1	1	0	1	On
0	1	0	0	0	0	0	0	1	1	0	1	On
1	1	0	0	0	0	1	0	0	1	1	0	Off
0	0	1	0	0	1	0	0	1	0	0	1	On
0	0	1	0	0	0	0	0	1	1	0	1	On
1	0	1	0	0	0	0	0	1	1	0	1	On
1	1	1	0	0	0	1	0	0	1	1	0	Off
0	0	0	0	1	0	0	0	1	0	0	1	On
1	0	0	1	0	0	0	0	1	1	0	1	On
1	0	1	1	0	0	0	1	0	1	1	0	Off
1	0	0	1	1	1	0	0	1	0	0	1	On
0	0	1	1	0	0	0	1	0	1	1	0	Off
1	0	1	0	1	0	0	0	1	0	0	1	On
1	0	1	0	0	0	0	0	1	1	0	1	On
1	1	1	1	0	0	1	1	0	1	1	0	Off

responses. The left six columns, A, B, C, D, E, and F, are the inputs. E and F are the reset inputs.

Notice that whenever the LED goes off, that step resulted in the set input to the latch (I) going active. The next step is the reset input, to return the circuit to the reset state. There are many conditions which will allow the circuit to set, and some which allow it to reset. Determine the conditions by analyzing the set and reset requirements.

Here again, there are some lines missing from the truth table. If desired, you can determine these by looking at the input conditions and complete the truth table.

After the circuit has been sequenced through its combinations, make up a signal loading chart from the specifications given and the example given in Chapter 6.

After this project is completed, you should try to design some of your own logic circuits. Draw the circuits out and analyze the operation before assembling. Make up the truth table, and then assemble and check the circuit.

PROJECT 11, 555 TIMER ONE SHOT

This project demonstrates the use of the 555 timer as a one shot. The LED provides a visual indication of the output pulse. The effect of changing the values in the timing circuit is demonstrated. The parts required are:

- 555 timer integrated circuit
- 2 1 K resistors, ¼ watt
- LED
- 47 K resistor
- 470 K resistor
- 10 μF capacitor
- 1 μF capacitor
- 0.01 μF capacitor
- Project 1, assembled

The schematic for this circuit is shown in Fig. 8-17, and the associated wire list is given in Table 8-20. This is a standard one-shot circuit with the timing components (R1 and C1) chosen to give about a ½ second delay. This can be checked by the equation for the delay time:

$$t = 1.1 \times R1 \times C1 = 1.1 \times 47\ K \times 10\ \mu F = 0.517\ sec$$

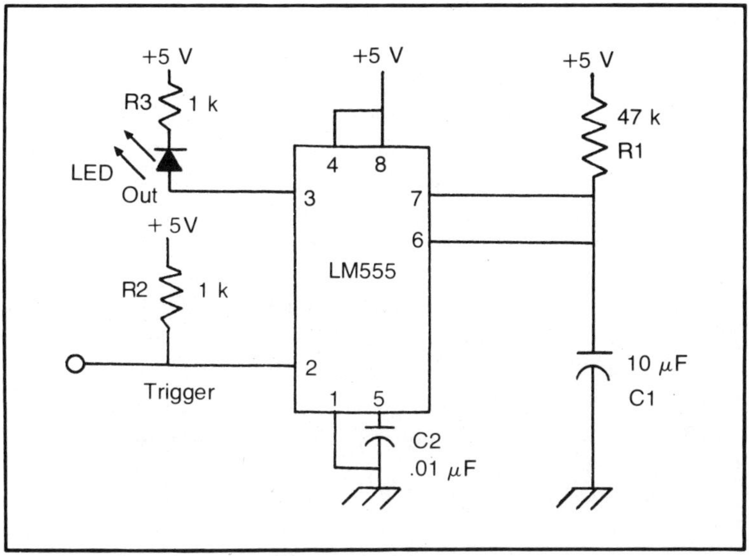

Fig. 8-17. The schematic for Project 11.

Assemble the circuit per the wire list, and carefully check the wiring. Then attach the battery clip to the battery. The LED should come on, indicating a low output. This is because the output is connected to act as a current sink, applying a ground to the negative side of the LED. Next momentarily touch the trigger input to ground. The LED should go off momentarily, then come back on. This indicates a short positive pulse (the time the LED was off) in response to the trigger input. The grounding of the trigger input is simulating a negative going pulse at the input. Trigger the input

Table 8-20. Wire List for Project 11

	R1	R2	R3	C1	C2	LED	IC
+5 Volts	Top	Top	Top				4,8
Ground				Low	Low		1
Trigger		Low					2
			Low			+	
						–	3
	Low			Top			6,7
					Top		5

several times to get a good indication of the off time of the LED.

Replace the 47 K resistor with the 470 K resistor, and trigger the circuit again, by grounding the trigger input. Notice that when the LED goes off, it stays off for a significantly longer time. The time-out time, and the pulse width for this condition is a factor of 10 greater than the previous circuit.

$$t = 1.1 \times 470\ \text{K} \times 10\ \mu\text{F} = 4.7\ \text{seconds}$$

Trigger the circuit several times to get a good indication of the time the LED is off. Then disconnect the battery and replace the 10 μF capacitor with the 1 μF capacitor. Reconnect the battery and trigger the circuit. Note that the LED goes off for a considerably shorter time than before. In fact, it should be the same as the time for the first trial.

The time-out time for the one shot is determined by the equation given above. If one of the components (R1) is increased by a factor of 10, the time will likewise be increased by a factor of 10. If one of the components (C1) is decreased by a factor of 10, the time will likewise decrease by a factor of 10.

The output of the timer is connected as a current sink so that the LED is on when the output is low. The 555 timer can also act as a current source to supply current to the load. To demonstrate this, reconnect the LED and the resistor (R3) to the output of the timer, as shown in Fig. 8-18. Now trigger the circuit. The LED comes on when the circuit is triggered, and is off before the trigger and after time out. Remember, when connecting the circuit, that the positive lead of the LED, which was connected to the resistor, is now connected to pin 3 of the timer. Also, the negative terminal of the LED is moved from pin 3 of the timer to the resistor.

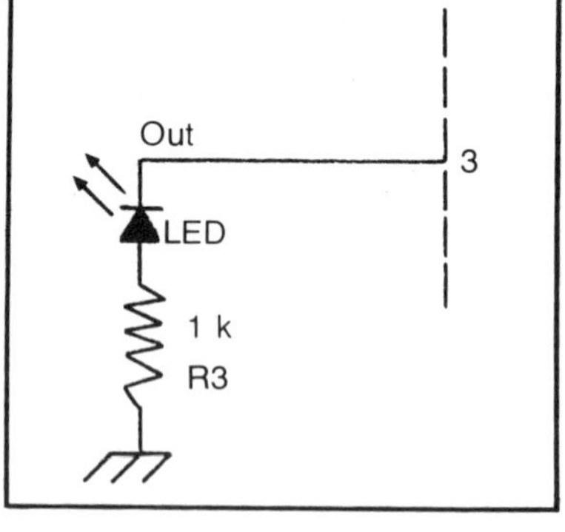

Fig. 8-18. Connecting the LED as a current sink load.

Remove the 1 K resistor, R2, which is biasing the trigger input high. Then touch your finger to a wire connected to the trigger input. Typically, the circuit will trigger erratically, indicating noise susceptibility of the trigger input. The bias resistor, R2, prevents the erratic triggering by holding this input high and not letting it float.

PROJECT 12, 555 TIMER FREE-RUNNING OSCILLATOR

This project demonstrates the use of the 555 timer integrated circuit as a free-running oscillator. It generates a square wave pulse train with a controllable frequency and duty cycle. The effect of changing the values of the timing components is demonstrated. The parts required are:

555 Timer
LED
2 1 K resistors
2 47 K resistors
470 K resistor
10 μF capacitor
1 μF capacitor
0.01 μF capacitor
Project 1, assembled

The schematic for this circuit is shown in Fig. 8-19, and the

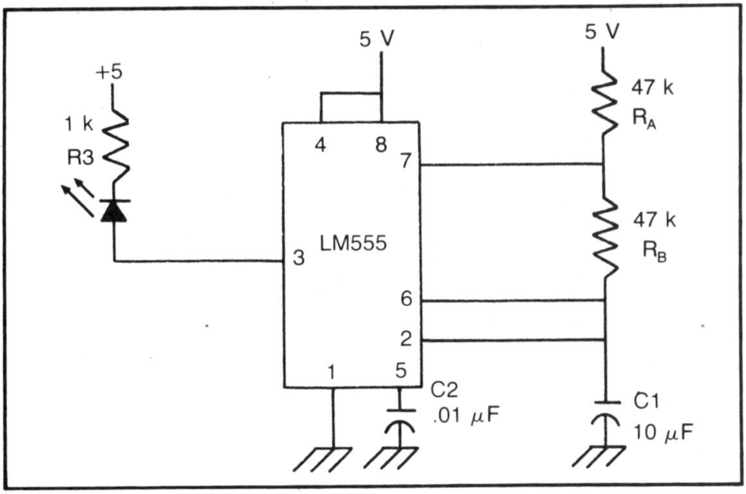

Fig. 8-19. The schematic for Project 12.

Table 8-21. Wire List for Project 12

	R_A	R_B	R3	C1	C2	LED	IC
+5 Volts	Top		Top				4,8
Ground				Low	Low		1
			Low			+	
						–	3
	Low	Top					7
		Low		Top			2,6
					Top		5

associated wire list is given in Table 8-21. This is the standard connection for a square wave generator. R_A, R_B, and C determine the time constants for the output waveshape.

Assemble the circuit and double check the wiring; then connect the battery. The LED should flash on and off at a little less than a one second rate with the off time being about twice the on time. The time can be determined by using the equations for t_{high} and t_{low}.

$$t_{high} = 0.693\ (R_A + R_B) \times C = 0.693(94\ K)\ 10\ \mu F$$
$$= 0.65 \text{ seconds}$$
$$t_{low} = 0.693\ (R_B)C = 0.693\ (47\ K)\ (10\ \mu F) = 0.325 \text{ seconds}$$

So the total cycle time is 0.975 seconds, and the duty cycle is:

$$D = \frac{R_B}{R_A + 2R_B} = \frac{47\ K}{47\ K + 94\ K} = ⅓ \text{ or 33 percent}$$

33 percent means that the output will be high (LED off) twice as long as the output will be low (LED on), as observed.

Now replace the resistor R_A with a 1 K resistor, and connect the battery. The LED on-time is the same, but LED off-time has been reduced to the almost on time.

$$D = \frac{47\ K}{1\ K + 94\ K} = 49.5 \text{ percent}$$

For this condition, the worst case current through the discharge pin

(pin 7) of the timer is 5 mA. The duty cycle can be made closer to 50 percent by decreasing R_A, but this also increases the current through the discharge pin.

Using the above equations for t_{high} and t_{low}, you can show that the times are almost equal.

Now replace R_A with the 470 K resistor, and reconnect the battery. The LED should sequence, with the off time several times longer than the on time. The on, or output low, time will be the same. This is because the low time, from the above equation, is a factor of R_B and C, and not of R_A. It will remain constant until one of these components is changed. The high time, or the time the LED is off, is:

$$t_{high} = 0.693(470\text{ K} + 47\text{ K})\ 10\ \mu\text{F} = 3.58\text{ seconds}$$

This is several times the 0.325 seconds for t_{low}, and is about what is observed.

The duty cycle for this condition is as follows:

$$D = 47\text{ K}/(470\text{ K} + 94\text{ K}) = 0.083 = 8.3\text{ percent}$$

Now, replace the 470 K resistor with the original 47 K resistor, and replace the 10 μF capacitor with the 1 μF capacitor. Examining the two timing equations shows that this is beyond the response frequency of the LED. It will appear to be on, but not at full intensity, and it may be flickering.

The reader can do the calculations to show that the total time is about 0.10 seconds, which is 10 cycles per second. The reader can also show that the duty cycle is the same as the first example (33 percent) and is not affected by the capacitor.

PROJECT 13, MC14066B ANALOG SWITCH

This project uses the MC14066B analog switch and the 555 timer to demonstrate the use of the analog switch in several different configurations. The parts required are:

- Project 1, assembled
- MC14066B analog switch integrated circuit
- 555 timer integrated circuit
- 47 K resistor, ¼ watt
- 3 1 K resistors, ¼ watt
- 10 μF capacitor

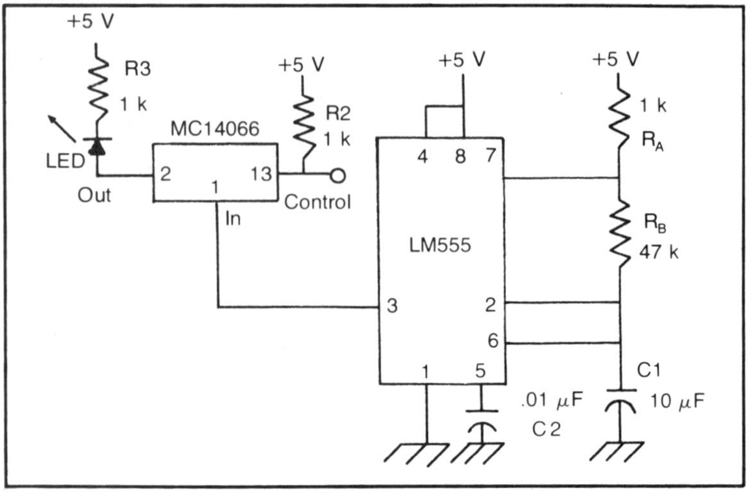

Fig. 8-20. The schematic for Project 13.

0.01 μF capacitor
LED

Figure 8-20 gives the schematic for this project, and the associated wire list is given in Table 8-22. The 555 timer is connected in the standard square wave generator configuration as in the last project. Review the previous project to determine the frequency of operation and the duty cycle. This square wave generator is used to supply a signal for the analog switch. The timer provides the signal

Table 8-22. Wire List for Project 13

	R	R_B	R2	R3	C1	C2	LED	LM555	MC14066
+5 V	Top		Top	Top				4,8	14
Gnd					Low	Low		1	7
Cont			Low						13
	Low	Top						7	
		Low			Top			2,6	
						Top		5	
								3	1
				Low			+		
							–		2

which is switched on and off by an external control. The LED indicates when the switch is on, by sequencing on and off.

Assemble the circuit according to the wire list. Double check the wiring to assure that it is correct and free from shorts. Connect the battery, the LED should sequence on and off. If it does not, connect the LED to pin 3 of the timer to make sure that this portion is working.

The control, in the initial condition, is biased high by resistor R2. Referring to the specifications for the analog switch, if the control input is high, the switch is closed, or on. If the control input is grounded, the switch is open, or off. Grounding the control input will turn off the switch, and the LED will become dark. Remove the ground and the LED will start sequencing again.

Since the circuits are powered by 5 volts, the control input to the analog switch can be controlled by a TTL logic gate. This is the next project, controlling the analog switch with logic circuits.

The above connection illustrated the use of the switch as a current sink carrying current from the load to the timer. To demonstrate that the analog switch can also act as a current source, connect the output of the analog switch as shown in Fig. 8-21. This configuration grounds the negative lead of the LED, through the limiting resistor. The drive current is supplied by the timer and gated by the analog switch. The operation of the switch is the same. A ground on the control input will open the switch, and the LED will be off. Take the ground off the control input and it will be biased high by R2. This closes the switch, and the LED sequences.

Figure 8-22 shows two additional circuit configurations. The upper circuit places the switch between the LED and the series limiting resistor. Here again, the circuit works in the same manner as above. When the control is high, the LED is sequencing indicat-

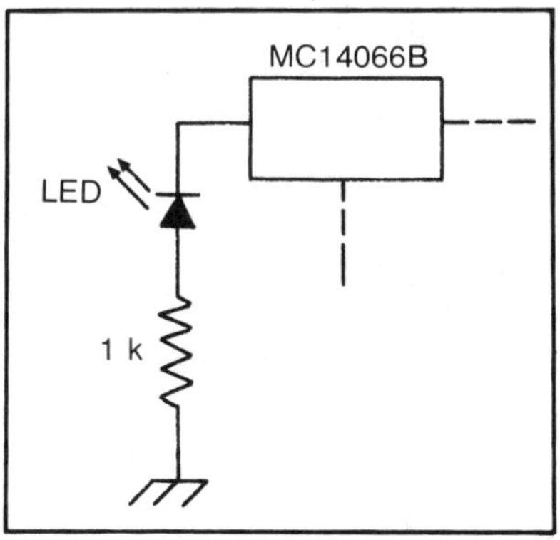

Fig. 8-21. Output connections for the analog switch to supply the LED current.

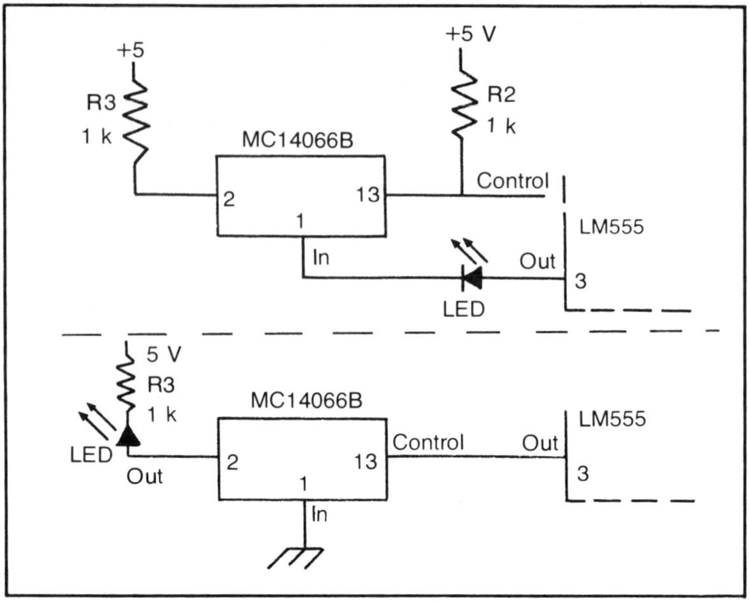

Fig. 8-22. Two additional analog switch configurations.

ing that the switch is closed. When the control input is grounded, the switch is opened, and the LED is off.

The lower part of Fig. 8-22 shows controlling the LED by turning off and on the switch with the square wave oscillator output. This switches the ground at the negative lead of the LED. For this circuit, the LED is on when the output of the 555 timer is high. This is because a high level turns on the switch and allows current to flow from the LED to ground. In the previous examples, the square wave from the 555 timer was used to turn the LED on and off. In the last example, the square wave turns the switch on and off, grounding the lower end of the LED. In this circuit, there is nothing magical about being between the LED and ground. It will work just as well if the switch is placed between R3 and+5 volts.

PROJECT 14, ANALOG SWITCH WITH TTL ICS

This project demonstrates the use of the analog switch to increase the current source and current sink capability of standard TTL digital logic circuits. The parts required are:

Project 1, assembled
MC14066B analog switch

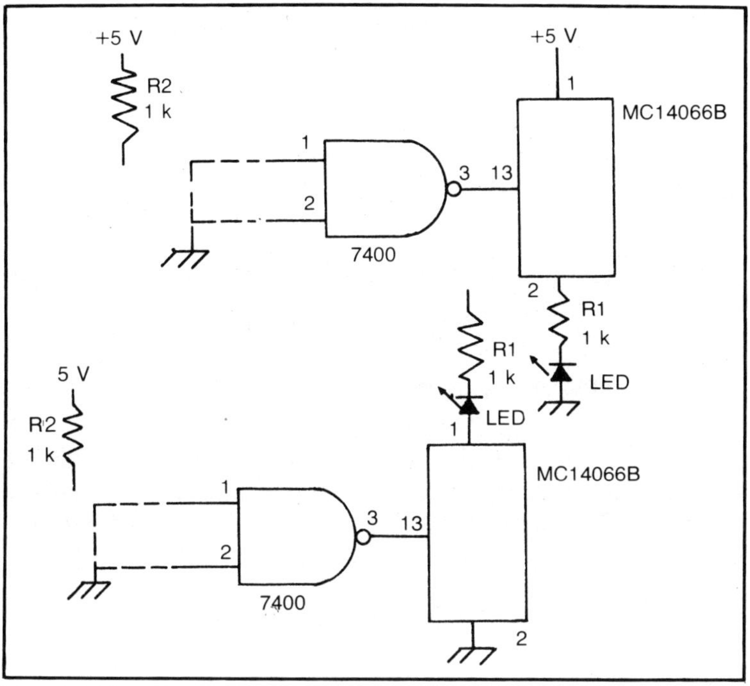

Fig. 8-23. The schematic for Project 14.

7400 integrated circuit
LED
2 1 K resistors, ¼ watt

The circuit schematics are given in Fig. 8-23. The wire list of Table 8-23 is for the upper circuit, which is to be assembled first. The analog switch, as wired, acts as a current source for the LED.

Table 8-23. Wire List for Project 14

	R1	R2	LED	7400	14066
5 Volts		Top		14	1,14
Ground			−	1,2,7	7
				3	13
	Top				2
	Low		+		

The TTL IC needs to supply only the signal required by the control input, and not the full LED current. Assemble the top circuit according to the wire list, and attach the battery. The LED should be on.

Remove both of the leads that are grounding the inputs to the NAND gate, and connect these to the lower end of the bias resistor. The LED should go off. Now, one at a time, connect both inputs to ground and notice that the LED comes on.

The lower circuit of Fig. 8-23 is a current sink type of circuit. Try to assemble and test this circuit on your own.

PROJECT 15, LM324 OP AMP AMPLIFIER

This project uses the LM324 operational amplifier as an audio amplifier to drive an LED. The input device for the amplifier is a speaker. The brightness of the LED is used as an indication of the amplitude of the output of the operational amplifier. This circuit is driven directly from the 9 volt battery. The parts required are:

LM324 operational amplifier
LED
3 1 K resistors
2 47 K resistors
2 27 K resistors
Speaker from an old transistor radio
9 volt battery and clip

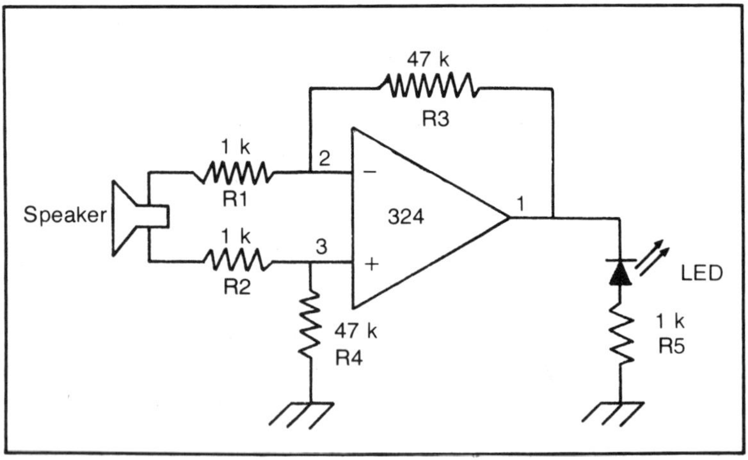

Fig. 8-24. The schematic for Project 15.

Table 8-24. Wire List for Project 15

	R1	R2	R3	R4	R5	LED	IC	Spkr
+5 Volts							4	
Gnd				Low	Low		11	
	Top		Low				2	
			Top			+	1	
	Low							Top
		Low						Low
		Top		Top			3	
					Top	–		

Figure 8-24 is the circuit schematic for this project, and Table 8-24 is the associated wire list. The operational amplifier is connected as a differential input amplifier. Both sides of the input are isolated from ground, and the impedance looking into the plus and the minus inputs are the same. In this circuit, the load requires current from the operational amplifier. It is a current source, driving the LED, in a variable brightness mode. The intensity of the LED is an indication of the amplitude of the output; the brighter the LED the higher the output.

Assemble the circuit according to the wire list and double check the wiring. When the wiring is correct, attach the battery to the battery clip; the LED should be off. Cup your hands around the speaker without touching it, and talk into the speaker. The LED should come on with varying intensity, depending how loud you talk into the speaker. Practice until full intensity is obtained. Humming into the speaker will tend to keep intensity more constant than words. Determine the volume required to just barely attain full intensity. Then replace the two 47 K resistors with 27 K resistors. Repeat the above process. Notice that it is harder, if not impossible, to attain full intensity of the LED. This is because the gain of the circuit has been cut in half by reducing the feedback resistors by two.

Try playing around with this circuit. Increase the gain and see what happens. Also, try connecting a speaker up to the output, as shown in Fig. 8-25. Then try talking into the input speaker and see if there is any output.

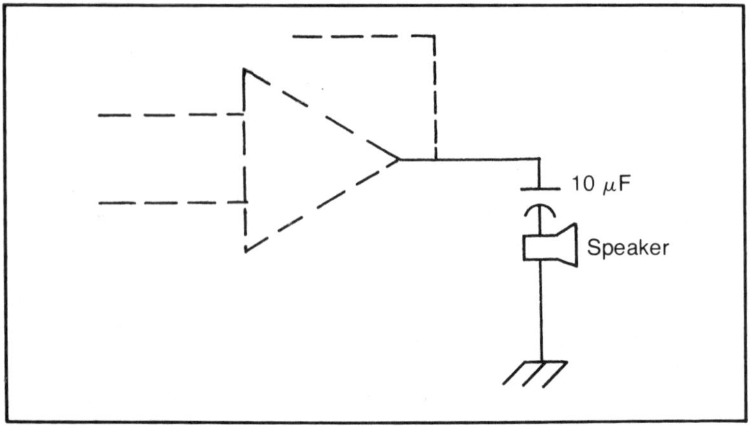

Fig. 8-25. Adding a speaker to the output of Project 15.

PROJECT 16, TWO STAGE LM324 AMPLIFIER

This project uses two of the operational amplifiers in the LM324 op amp in a two-stage amplifier. The first stage is identical to the previous project. The second stage is a noninverting amplifier. The parts required for this project are:

LM324 operational amplifier
LED
4 1 K resistors
3 47 K resistors
Speaker from an old transistor radio
9 volt battery and clip

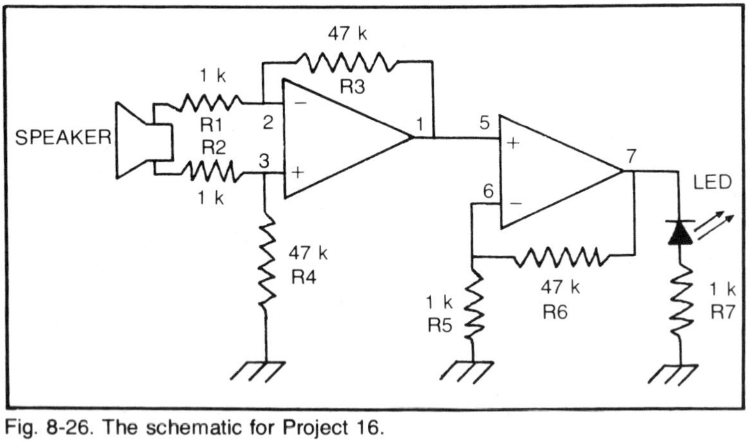

Fig. 8-26. The schematic for Project 16.

Table 8-25. Wire List for Project 16

	R1	R2	R3	R4	R5	R6	R7	LED	IC	Spkr
+5 V									4	
Gnd				Low	Low		Low		11	
	Top		Low						2	
	Low									Top
			Top						1,5	
		Low								Low
		Top		Top					3	
					Top	Low			6	
						Top		+	7	
							Top	−		

The schematic for this circuit is shown in Fig. 8-26, and the associated wire list is shown in Table 8-25. This two-stage amplifier has a gain of 47 for each stage and a combined gain of 47 times 47, or 2209. Again, the LED is used as an output indication and the speaker as the input.

Assemble the circuit per the wire list according to Table 8-25 and double check the wiring. When you connect the battery the LED should be off. Cup your hands around the speaker without touching it, and talk into the speaker. It takes considerably less volume to achieve full intensity of the LED. This is because of the increased gain.

Set the input speaker close to the speaker of a radio. Adjust the radio volume and the distance between the speakers until the LED comes on full intensity at the peaks in volume and the intensity varies considerably. This is a visual indication of what is coming out the radio speaker. Connect the second speaker, as shown in Fig. 8-25, to the output in place of the LED. The volume from the speaker will be low, but audible. There will be considerable distortion present.

You can also try replacing the input speaker with a microphone and changing the gain of the stages.

PROJECT 17, 7474 FLIP-FLOP

This project demonstrates the function of the set and reset inputs for the 7474 flip-flop and the use of this flip-flop as a set/reset flip-flop. The parts required for this project are:

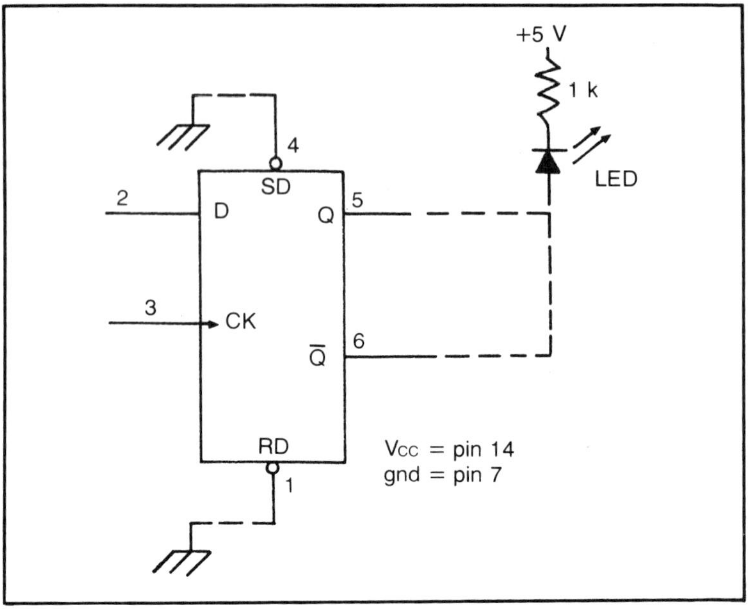

Fig. 8-27. Schematic for Project 17.

Project 1, assembled
7474 integrated circuit
LED
1 K resistor, ¼ watt

The schematic for this circuit is given in Fig. 8-27, and the associated wire list is given in Table 8-26. Assemble the circuit as shown in the wire list and connect the battery. The LED should be on because it is connected to the inverting output, and the set input (SD) is grounded. Connect the LED to the noninverting output (pin 5), and the LED should be out. This is the set condition, with the noninverting output high and the inverting output low.

	R	LED	7474
5 Volts	Top		14
Ground			7, 4
	Low	+	
		−	6

Table 8-26. Wire List for Project 17

Remove the ground from the set input (pin 4) and ground the reset input (pin 1). The LED should be on. Connect the LED to the inverting output (pin 6). It should be off. This is the reset condition of the flip-flop with the inverting output high and the noninverting output low.

The set and reset inputs of the flip-flop are used to establish a known condition for the flip-flop. Notice that once the condition is established, the condition will not change when the ground is removed from the set or reset input. Try it by grounding the set input and determining the flip-flop output conditions, then remove the ground. The condition will remain the same. Try the same thing with the reset input. This capability of the flip-flop to remember the last commanded condition (or state) is used in computer and logic circuits.

PROJECT 18, 7474 COUNTDOWN CIRCUIT

Flip-flops have the unique characteristic of being able to change states once for every two changes of state of the input. This gives the flip-flop the capability of dividing the input pulse rate by 2, or of making the output square wave twice the time period of the input. This ability is the function of the basic countdown circuit.

The parts required for this project are:

Project 1, assembled
Project 12, assembled
7474 integrated circuit
LED
1 K resistor

The schematic for this project is given in Fig. 8-28, and the associated wire list is given in Table 8-27. The square wave generator from Project 12 is used to provide the input pulses. The power source for the circuit is Project 1. The resistor and LED on the input to the first stage are components which are a part of Project 12. Use the resistors and capacitor from Project 12 which gave a ½ to 1 second on and off time for the LED. This gives a LED indication which is easily distinguishable. Use the LED and resistor connected to the output of Project 12 to monitor the pulse rate.

To use the flip-flop as a countdown circuit, one of the outputs must be connected back to the input to enable the circuit to change states on the next clock pulse input. Referring to the mode select table for this IC, the D input and the inverting output are compatible

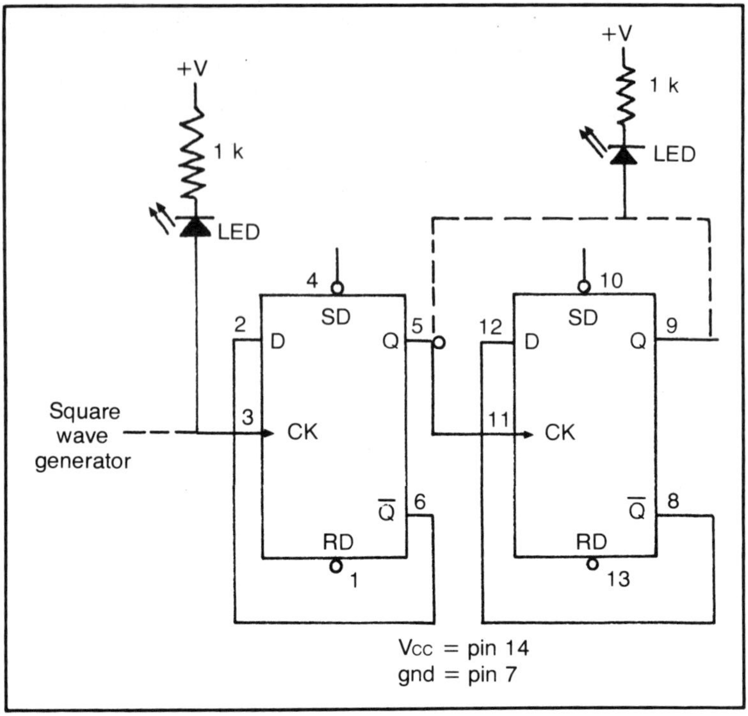

Fig. 8-28. Schematic for Project 18.

because the previous state requirements for the D input are the opposite those of the current inverting output. So feeding the inverting output back into the D input enables the circuit to change states on the clock pulse input.

	R	LED	IC Pin
5 Volts	Top		14
Ground			7
Input			3
			2,6
			5,11
			8,12
	Low	+	
		−	9

Table 8-27. Wire List for Project 18

Assemble the circuit according to the wire list. Assemble the pulse generator first, and verify that it is working before connecting the +5 volts to the 7474 IC. When the pulse generator is sequencing, as indicated by the LED on the output of the pulse generator, connect power to the 7474. The output LED will be sequencing at ¼ the rate of the LED connected to the pulse generator output. Connect the LED to the other output of the second stage (pin 8), note that the pulse rate is the same. Next, connect the LED to the noninverting output of the first stage (pin 5). The LED sequences at ½ the rate of the input. The circuit consists of two divide-by-two circuits. Each stage divides its input by two so that this circuit divides the input pulse rate by 4.

Next, connect the LED to the noninverting output of the second stage (pin 9) and ground the reset pin for the second stage (pin 13). This will force the output to the reset condition with the noninverting output low and the LED on. Notice that the sequencing for the second stage has stopped. The LED connected to pin 9 is on continuously instead of sequencing on and off. Remove the ground from pin 13 and the LED will start to sequence again. Ground the set pin for the second stage (pin 10) and the LED will go off and stay off. The second stage is in the set condition with the noninverting output high. Connect the LED to the inverting output (pin 8). It will be on.

Ground the reset input for the first stage (pin 1). The LED will stop sequencing and may be on or off depending on the condition of the second stage when the ground was applied. Remove and reapply the ground a few times until both conditions (LED on and LED off) are achieved. Connect the LED to pins 5 and 6 (the noninverting output and the inverting output of stage 1) to verify that when pin 1 is grounded, the first stage is in the reset condition. In a counter of more than one stage, when one stage is held in the set or reset condition, the following stages stop sequencing. This is because the input is no longer changing states.

Remove the ground from pin 1 and ground pin 4 (the set input for the first stage). Verify that the first stage is in the set condition and that the second stage has stopped sequencing. Use the LED for this.

PROJECT 19, FOUR STAGE COUNTDOWN CIRCUIT

This circuit uses two 7474 flip-flops in a four stage countdown circuit. It will divide the input pulse rate by a factor of 16. In addition, logic gates are used to generate pulses for some of the

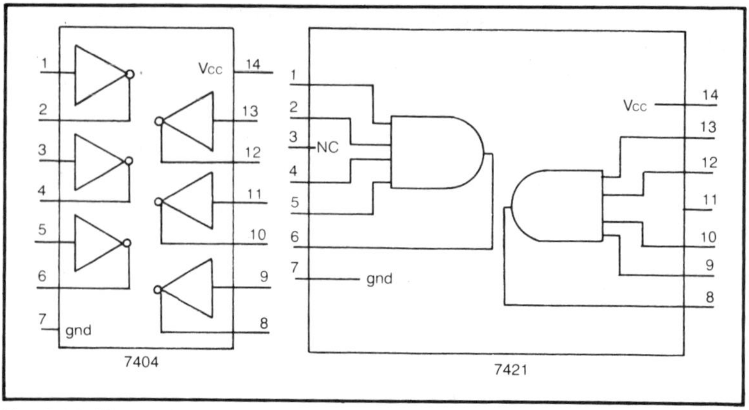

Fig. 8-29. Pin configuration for the 7404 hex inverter and the 7421 dual 4-input AND gate.

individual time periods. This requires two ICs which have not been specifically discussed, but they have been described in general. These are the 7404 hex inverter and the 7421 dual 4-input AND gate. The function of these type of gates has been covered, and the

Table 8-28. Wire List for Project 19

	R	LED	IC1 pin	IC2 Pin
5 Volts	Top		14	14
Ground			7	7
Input			3	
	Low	+		
		−		9
			2,6	
			5,11	
			8,12	
			9	3
				2,6
				5,11
				8,12

pin configuration and logic diagrams for these ICs are shown in Fig. 8-29.

The parts required for this project are:

Project 1, assembled
Project 12, assembled
2 7474 Dual D-type flip-flops
7404 hex inverter
7421 dual 4-input AND gate
3 LEDs
3 1 K resistors

The schematic for the 4 stage countdown circuit is shown in Fig. 8-30. The associated wire list is given in Table 8-28. The pulse generator of Project 12 is used to generate the input pulse train. Select the timing components for Project 12 such that the output LED of the generator will sequence about once a half second to one second. This will make observation easier, and provide definite indications of the state of the signal.

Note that this project and the schematic of Fig. 8-30 essentially consists of two of the circuits of project 18 connected in series. This can be carried out to several stages, involving long time periods.

Wire the circuit according to the wire list. Double check the wiring, and look for shorted wires. Disconnect the +5 volts to both 7474 ICs and connect the power source. This will apply power to the square wave generator. When you are assured that the generator is sequencing, connect the +5 volts to the 7474 ICs. When the power source is reconnected, the output LED will start to sequence at 1/16 the rate of the input LED. This will be once about every 8 to 16 seconds, so do not get impatient.

If there is doubt about the circuit sequencing, make sure the input LED is sequencing. Then, starting with the output of the first stage, connect the LED to the stage outputs, one at a time to determine which stages are sequencing. The problem will be in the first stage whose output is not sequencing. Using the schematic for reference, check the circuitry associated with this stage. Make sure that the +5 volts and ground are connected securely to the IC. Check that the wiring is going to the proper pin numbers and that there are no shorts. If the LED connected to the output of the last stage is sequencing properly, this is a good indication that the circuit is working properly.

When the circuit is working properly, connect the LED to all

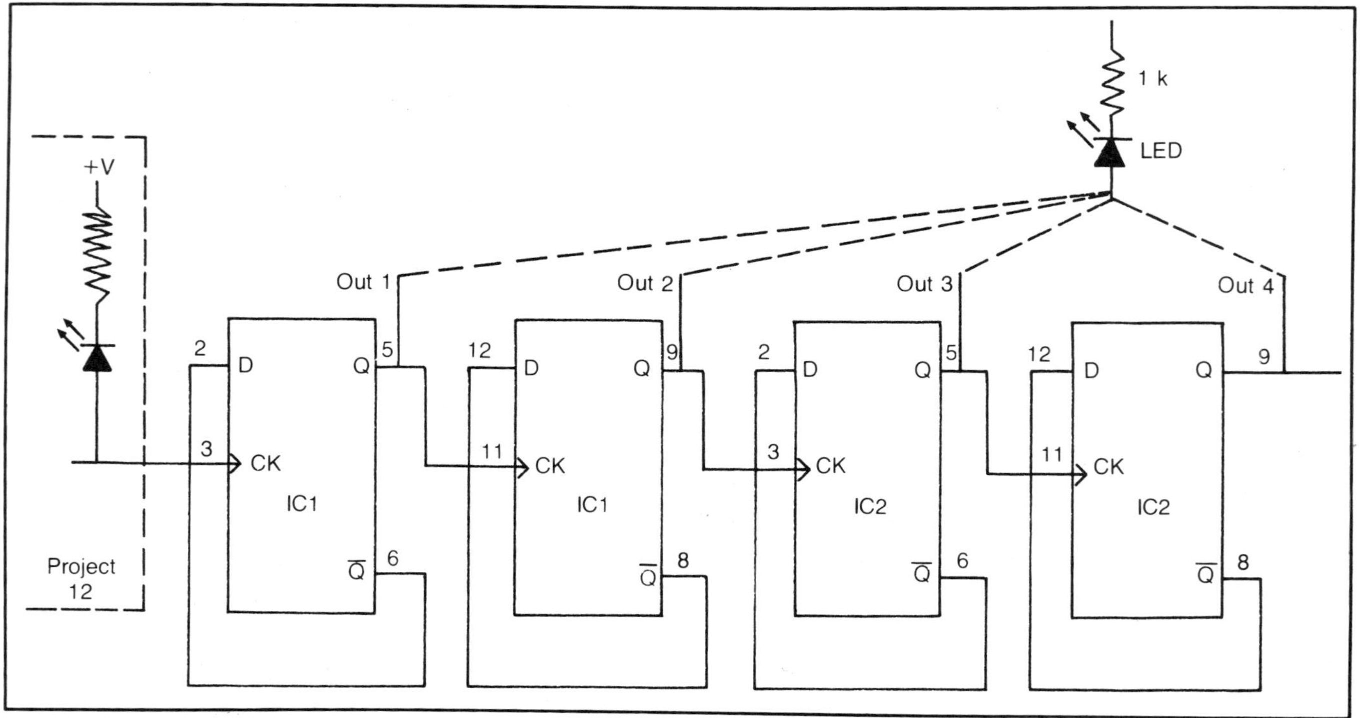

Fig. 8-30. Schematic for Project 19.

four output pins, one at a time. Notice that the output pulse rate is ½ that of the output of the preceding stage. The pin numbers for the ICs are shown here in order of lengthening pulse rate.

7474 #1, Pin 2—Input pulse train
7474 #1, Pin 5—Output 1
7474 #1, Pin 9—Output 2
7474 #2, Pin 5—Output 3
7474 #2, Pin 9—Output 4

Leave the LED connected to the output of the last stage. This LED will indicate the slowest pulse rate.

If one of the above steps fails to produce a sequencing rate of ½ that of the previous stage, check the wiring and correct any errors found. If there are no errors in the wiring, and there is no indication of the problem, it is always possible that the IC is bad. Occasionally an IC will get damaged due to circuit problems, or mishandling.

When this portion of the circuit is working correctly, add the hex inverter and the 4-input AND gate to generate the individual pulses in the various time periods. The hex inverter is used to generate the NOT signals required, and the 4-input AND gate is used to logically AND the desired signals to produce the desired pulse. Table 8-29 gives the connections for the inverter and the AND gate to produce all the 16 possible combinations of input signals. This gives the logic requirements to generate any one of the 16 possible time period pulses available. If there are no pin numbers in the column for the inverter, the signal does not require an inverter, and is connected directly to the AND gate.

Note that the first time period is for time 0, or before the first pulse after the circuit has counted full and started over. The output of the first counter section is connected to an inverter input. The corresponding inverter output is connected to the AND gate input, generating a NOT signal input to the AND gate. The output of the AND gate is connected to the input of another inverter section. The output of the inverter in the right hand column is connected to an LED and series limiting resistor to +5 volts. This LED connection and series limiting resistor are not given in the table, but remember that an LED will indicate the logic state. Always use a series limiting resistor between the positive lead of the LED and +5 volts. Also, remember that +5 volts must be connected to pin 14 and ground to pin 7 of the 7421 and the 7414.

In the table, there are four lines associated with each time

Table 8-29. Inverter and AND Gate Connections for Project 19

Time Period	Counter Output	Inverter		AND Gate		Inverter	
		in	out	in	out	in	out
0	Out 1	1	2	1	6	11	10
	Out 2	3	4	2			
	Out 3	5	6	4			
	Out 4	9	8	5			
1	Out 1			1	6	11	10
	Out 2	3	4	2			
	Out 3	5	6	4			
	Out 4	9	8	5			
2	Out 1	1	2	1	6	11	10
	Out 2			2			
	Out 3	5	6	4			
	Out 4	9	8	5			
3	Out 1			1	6	11	10
	Out 2			2			
	Out 3	5	6	4			
	Out 4	9	8	5			
4	Out 1	1	2	1	6	11	10
	Out 2	3	4	2			
	Out 3			4			
	Out 4	9	8	5			
5	Out 1			1	6	11	10
	Out 2	3	4	2			
	Out 3			4			
	Out 4	9	8	5			
6	Out 1	1	2	1	6	11	10
	Out 2			2			
	Out 3			4			
	Out 4	9	8	5			
7	Out 1			1	6	11	10
	Out 2			2			
	Out 3			4			
	Out 5	9	8	5			
8	Out 1	1	2	1	6	11	10
	Out 2	3	4	2			
	Out 3	5	6	4			
	Out 4			5			
9	Out 1			1	6	11	10
	Out 2	3	4	2			
	Out 3	5	6	4			
	Out 4			5			

Table 8-29. Inverter and AND Gate Connections for Project 19 **(Continued from page 242)**

Time Period	Counter Output	Inverter		AND Gate		Inverter	
		in	out	in	out	in	out
10	Out 1	1	2	1	6	11	10
	Out 2			2			
	Out 3	5	6	4			
	Out 4			5			
11	Out 1			1	6	11	10
	Out 2			2			
	Out 3	5	6	4			
	Out 4			5			
12	Out 1	1	2	1	6	11	10
	Out 2	3	4	2			
	Out 3			4			
	Out 4			5			
13	Out 1			1	6	11	10
	Out 2	3	4	2			
	Out 3			4			
	Out 4			5			
14	Out 1	1	2	1	6	11	10
	Out 2			2			
	Out 3			4			
	Out 4			5			
15	Out 1			9	8	13	12
	Out 2			10			
	Out 3			12			
	Out 4			13			

period, one line for each of the counter output signals. Starting with the counter output, each pair of numbers on a line represents one wire or one connection. The left hand number represents the source of the signal and the right hand number represents the destination. The last, or right hand, column designates the inverter output pin number to be connected to the negative lead of the LED monitoring the time pulse. Most of the time periods call for inverter pin 10 to be connected to the LED, but the last time period calls for inverter pin 12 to be connected to the LED. This is so that the LED monitor can be left connected to the last time period.

As an example, let us connect an LED monitor for the last time period, when all the outputs are high. From the table it says to make the following connections:

Counter output 1 to AND gate pin 8

AND gate pin 8 to inverter pin 13
Inverter pin 12 to LED negative lead
LED positive lead to series resistor low (from text)
Series resistor high to + 5 volts (from text)
Counter output 2 to AND gate pin 10
Counter output 3 to AND gate pin 12
Counter output 4 to AND gate pin 13

Connect the above connections and connect power to the circuit. The LED at the inverter output will be on during this time period. Connect the remaining LED to output 1 of the counter. The time duration of the pulse from the AND gate should be the same as that of the LED connected to output 1. But the AND gate output will sequence once while the output of counter 1 will sequence 8 times. Leave this portion of the AND gate connected to indicate when the count has reached full.

Next, connect the remaining gates according to the first line of the table. Connect the LED and series resistor, previously connected to output 1, to the output of the inverter (pin 10). This LED is connected to monitor time period 0, which immediately follows time period 15. The other LED is connected to monitor time period 15. The LED monitor time period 0 should sequence immediately after the LED monitoring time period 15.

Refer to the waveshapes in Chapter 6 to clarify when the time periods are generated. When this circuit is working correctly and understood, connect the AND gate as shown in another line of the table. Determine if the time period indicated is the desired time period by judging the time that the LED is on in relation to the time period 15 LED. This should be repeated for at least 4 more time periods in the table. Always observe and understand the indications. As usual, disconnect the power source before changing any of the connections.

PROJECT 20, 7485 4-BIT MAGNITUDE COMPARATOR

This project demonstrates the use of the 7485 comparator to compare the magnitude of two 4-bit digital words. The two binary words are A and B, and the individual bits are A3, A2, A1, and A0 for word A, and B3, B2, B1, and B0 for word B. Bits 3 are the most significant bits and bits 0 are the lowest order bits. There are three outputs, $A < B$, $A = B$, and $A > B$, which indicate the results of the compare operation.

The parts required for this project are:

Project 1, assembled
7485 4-bit magnitude comparator IC
3 LEDs
4 1 K resistors

The schematic for this project is given in Fig. 8-31, and the associated wire list is given in Table 8-30. If three LEDs are not available, one can be used by connecting it to the various outputs, one at a time.

Assemble the circuit according to the wire list. After the wiring has been checked, connect the power source. Table 8-31 gives the output indications for the various input conditions. An X in the table means that that input has no affect on the output because of the previous input conditions on that line. Remember that the LED is on for a lcw condition, and it is off for a high, or true indication.

The circuit, as wired, should come up with the indications given on line 1 of the table. This is because all the inputs are open, and this is usually interpreted as a high condition for TTL logic

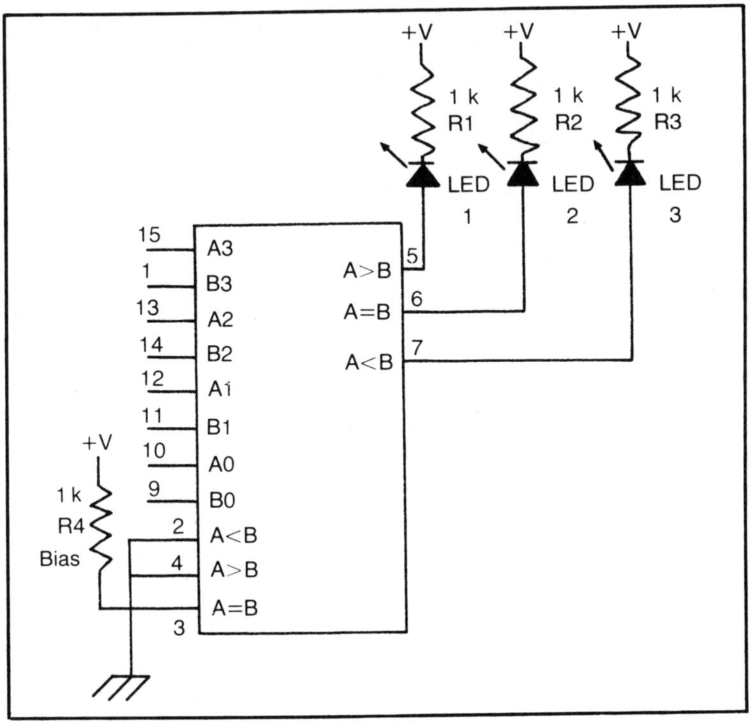

Fig. 8-31. Schematic for Project 20.

Table 8-30. Wire List for Project 20

	R1	LED1	R2	LED2	R3	LED3	R4	IC pin
5 volts	top		top		top		top	16
ground								2, 4, 8
	low	+						
		−						5
			low	+				
				−				6
					low	+		
						−		7
							low	3

circuits. If the indications are not as specified, connect all the inputs to the lower end of the bias resistor (R4). When this is correct, sequence through the table, one line at a time. Make sure the LED indications are as indicated. connect the Xs in the table input to ground and then to the bias resistor. The output indications should

Table 8-31. Sequencing Table for Project 20

A3	B3	A2	B2	A1	B1	A0	B0	Outputs - LEDs		
15	1	13	14	12	11	10	9	A>B	A<B	A=B
H	H	H	H	H	H	H	H	On	On	Off
H	L	X	X	X	X	X	X	Off	On	On
L	H	X	X	X	X	X	X	On	Off	On
L	L	H	L	X	X	X	X	Off	On	On
L	L	L	H	X	X	X	X	On	Off	On
L	L	L	L	H	L	X	X	Off	On	On
L	L	L	L	L	H	X	X	On	Off	On
L	L	L	L	L	L	H	L	Off	On	On
L	L	L	L	L	L	L	H	On	Off	On
L	L	L	L	L	L	L	L	On	On	Off

not change. For the L, or low inputs, connect them to ground, and for the H or high inputs, connect them to the bias resistor.

Table 8-30 sequences the inputs through the comparing inputs and does not take into account the cascading inputs. The cascade inputs are connected to simulate an A = B input. Since the cascading inputs are active high, the $I_{A=B}$ input is connected to the bias resistor and the other two cascading inputs are grounded. To demonstrate the operation of the cascading inputs, connect all the A and B inputs to ground. The A = B LED should be off because all the inputs are equal. Next, remove the ground from the $I_{A<B}$ input (pin 2), and connect it to the bias resistor. Disconnect the $I_{A=B}$ input (pin 3) from the bias resistor and connect it to ground. This simulates an A<B input into the cascade inputs. The A<B LED should be off, and the other two LEDs should be on.

To simulate the A>B cascading input, remove pin 2 from the bias resistor and connect it to ground. Remove the ground from the $I_{A>B}$ input (pin 4) and connect it to the bias resistor. The A>B output LED should be off and the other two LEDs on, indicating an A>B condition.

To simulate the A=B cascading input, remove pin 4 from the bias resistor and connect it to ground. Then remove the $I_{A=B}$ input (pin 3) from ground and connect it to the bias resistor. The A=B output LED should be off, and the other two LEDs on, indicating an A=B condition.

PROJECT 21, SEVEN-SEGMENT DISPLAY

The seven-segment LED display is a common numeric display used in conjunction with logic circuits. It displays the decimal digits 0 thru 9 by dividing the numbers into segments. This project is to familiarize the reader with the use of seven-segment displays by connecting one in a simple circuit and selecting the individual segments to make them illuminate.

The parts required for this project are:

Project 1, assembled
Seven-segment display (see text)
7 1 K resistors

The seven segment display used in this project is the HP5082-7730. But if that display is not available, any common anode display which fits into a standard DIP socket will work. Check the pin configuration for the display you use against that for the HP5082-7730, as

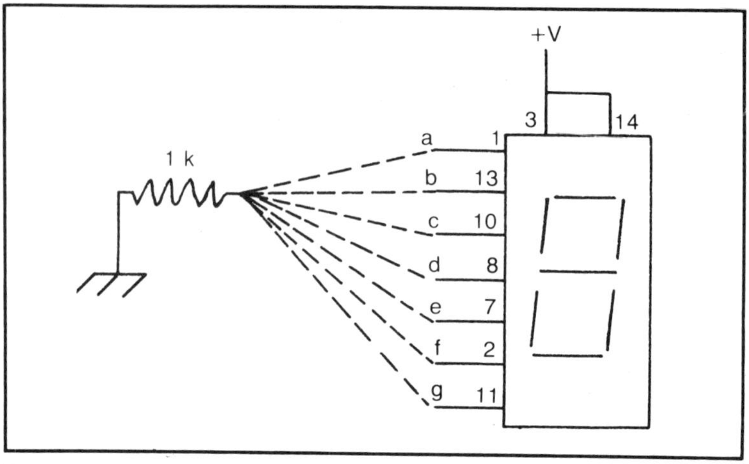

Fig. 8-32. Schematic for Project 21.

given in Chapter 6. If they are the same, proceed. If there is a difference, change the pin numbers in the project to correspond to those of the display used. Refer to the discussion about displays in Chapter 6.

Figure 8-32 is the schematic for this project, and the associated wire list is shown in Table 8-32. Assemble the project according to the wire list. Do not connect any of the segment pins yet. Connect the power source. No segments should be on. Next, connect the segments through a 1 K, resistor to ground, one segment at a time.

As each of the segment pins is connected through the resistor to ground, it should light. Notice where the individual segments are located on the display.

Now that the individual segments have been checked out, connect one resistor from the segment b pin to ground, and another resistor between the segment c pin to ground. Segments b and c should illuminate, giving the number 1 on the display. Using the segment table in Chapter 6, connect resistors between the various segment pins and ground to form the numbers. **Never, never connect a segment pin directly to ground.** This will destroy that segment.

	R	Display
+5 Volts		3, 14
Ground	Low	

Table 8-32. Wire List for Project 21

A display can be used if one of the segments is burned out as long as it is realized that the numbers will not be complete.

PROJECT 22, 7490 DECADE COUNTER

The 7490 integrated circuit is a decade counter which gives a BCD indication of the number of counts received. This IC is used extensively in decimal counter circuits, and in particular, those with LED indicators. But this project looks only at the 7490 counter, and determines the outputs for the various count inputs.

The parts required for this project are:

Project 1, assembled
Project 12, assembled
7490 IC
LED
1 K resistor

The schematic for this project is given in Fig. 8-33, and the associated wire list is given in Table 8-33. Here again, the square wave generator of Project 12 is used as the input signal generator. Select the timing resistors and capacitor for Project 12 to give a pulse rate of between ½ second and 1 second.

The 7490 is essentially a four stage count down circuit with the resets connected so that the circuit resets to zero when the count of 9 is reached. This gives a full decade count, from 0 to 9, and starting

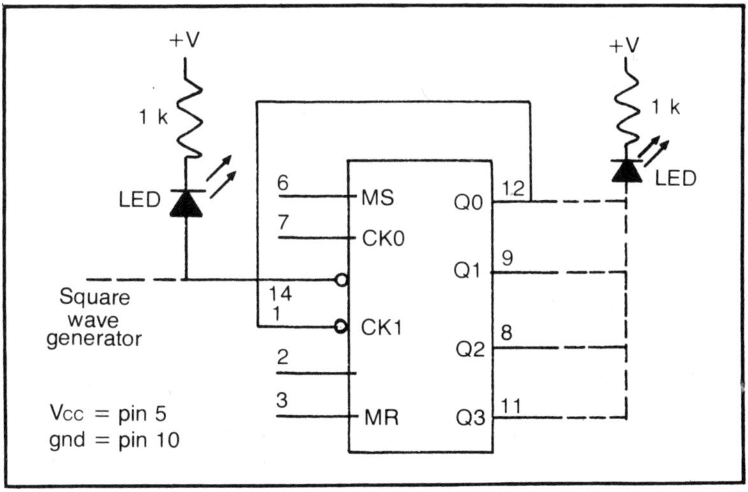

Fig. 3-33. Schematic for Project 22.

	R	LED	IC Pin
+5 Volts	Top		5
Ground			2,3,6,7,10
	Low	+	
		–	12,1
Input			14

Table 8-33. Wire List for Project 22

over at 0. The IC has a master set and reset to control the starting count, and to inhibit counting. Refer to Chapter 5 and 6 for a discussion on the operation of the set and reset inputs.

Assemble the circuit according to the wire list and connect the power source. The LED is connected to the low order count, Q0, and gives an indication of this stage output. The LED will sequence at ½ the rate of the input, or the output LED sequences once for every two input pulses. When this indication is correct and understood, connect the LED to the next higher output, Q1 (pin 9). Here the LED should be on for two counts of the input, and off for two counts, or sequence once for every 4 input pulses. Next connect the LED to the Q2 output (pin 8). Here the LED is on for four input pulses and off for four input pulses, or it sequences once for every 8 input pulses.

Next, connect the LED to the remaining output, Q3 (pin 11). Here the LED is on for eight input counts, and off for two counts. It sequences once for every 10 input pulses. This output is the only output which is not symmetrical, and has a shorter off time than on time. This is because the internal circuitry of the IC resets the circuit when a count of 10 is reached.

Remove the grounds from MR1 and MR2 (pins 2 and 3). Connect the LED to the output pins, one at a time (pins 12, 9, 8, and 11). Notice that the count has stopped and all the outputs are low (LED on). This is the reset condition for the circuit. Next ground pins 2 and 3 and remove the grounds from MS1 and MS2 (pins 6 and 7). Again, connect the LED to the output pins, one at a time. Notice that the count has again stopped, but that pins 12 and 11 are high. This is the set condition of the circuit, and is the count of 9 from the BCD count table. It may be necessary to use a bias resistor connected to+5 volts and to connect the set and reset inputs to the lower end. Normally this is not required.

PROJECT 23, MICROPHONE AMPLIFIER

This project uses the MC1306 power amplifier IC in a simplified microphone amplifier circuit. The preamplifier gain is 100, and the gain of the second stage is 10. The overall gain is 1000. A volume control is provided by a variable resistor connected to the input of the preamplifier. For this to be a useful circuit, or anything besides a circuit to play with, additional work must be done on the circuit. Such things as providing a filter circuit on the output to prevent oscillations, splitting the series resistor for the power stage, and providing a capacitor for stability will enhance the circuit.

The parts required for this project are:

9 volt battery, with battery clip
MC1306 power amplifier
Microphone from cassette recorder
10 K variable resistor
10 K resistor
1 megohm resistor
1 K resistor
2.7 ohm resistor
2 0.1 μF capacitors
15 pF capacitor
0.05 μF capacitor
200 μF capacitor, at least 25 volts
Speaker (at least 8 ohms)

The schematic for the microphone amplifier is given in Fig. 8-34, and the associated wire list is given in Table 8-34. R2 is the series resistor, and R3 is the feedback resistor for the preamplifier stage. The value of these two resistors sets the preamplifier gain. Dividing 1 megohm by 10 K gives 1000, which is the preamplifier gain.

R5 is to limit the ac output current. The volume control is R1.

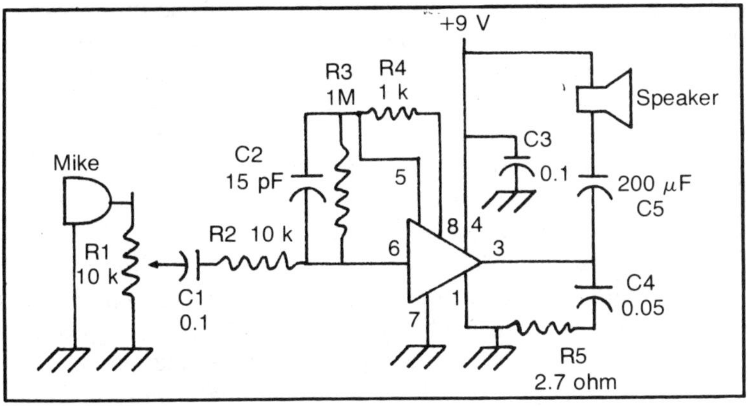

Fig. 8-34. Schematic for Project 24.

This variable resistor acts like a voltage divider for the microphone input, thus controlling the volume. For the microphone, use the remote microphone from a portable cassette recorder. For the speaker, use a speaker from an old transistor radio. It must have at least 8 ohms impedance. The speaker impedance is usually marked on the back of the voice coil. If a speaker of less than 8 ohms is used, there is the possibility of damaging the MC1306 IC by drawing excessive current. If a pair of low impedance headphones are available, they can be used in place of the speaker.

Wire the circuit according to the wirelist. Connect the microphone into the circuit using alligator clips or clip leads. Do not cut the wire from the microphone to the plug unless the microphone is not to be used with the recorder. Use the clip leads, and clip onto the plug with one end, and into the circuit with the other. The microphone plug has two sections separated by a black washer. These two sections are the two connections to be made to the microphone. Some microphones have a dual plug. One plug is the pause input to the recorder from a switch on the microphone. Make sure to select the proper plug if there are two.

Solder leads to the speaker terminals to connect it into the circuit. Connect all the grounds to terminals on the test board which are in the same area, preferably adjacent. This will act as a single point ground, reducing the coupling. When all the connections are made, double check the wiring for errors and shorts. When everything looks good, adjust the volume control to about the midpoint and connect the battery. Listen to the speaker, there should be a hiss. If a tone or squeal is heard, this is an indication that the circuit is oscillating, or that the microphone is too close to the speaker. Always point the speaker away from the microphone and separate them by a couple feet. This will reduce the tendency of the speaker feeding back into the microphone. If this does not cure the problem, carefully separate the wires in the circuit as far as possible to reduce coupling between them. It may be necessary to reduce the volume.

If no sound is heard in the speaker, try tapping the microphone. If this is not heard in the speaker, increase the volume control. If you still do not hear anything, double check the wiring.

A low level hissing sound is to be expected in the average circuit. This is a result of the noise which enters the circuit as a result of the test board and test leads. When the circuit appears to be working as well as can be expected, try talking into the microphone. The sound from the speaker may be distorted, but it should be understandable. Lowering the volume may reduce the distortion by reducing the drive to the preamplifier.

Table 8-34. Wire List for Project 23.

	Mike	R1	C1	R2	C2	R3	R4	R5	C3	C4	C5	Spkr	IC pin
+5 volts									+			top	4
Ground	low	low						low	–				1, 7
	top	top											
		CT	–										
			+	top									
				low	–	low							6
					+	top	top						5
							low						8
										+	–		3
								top		–			
											+	low	

9

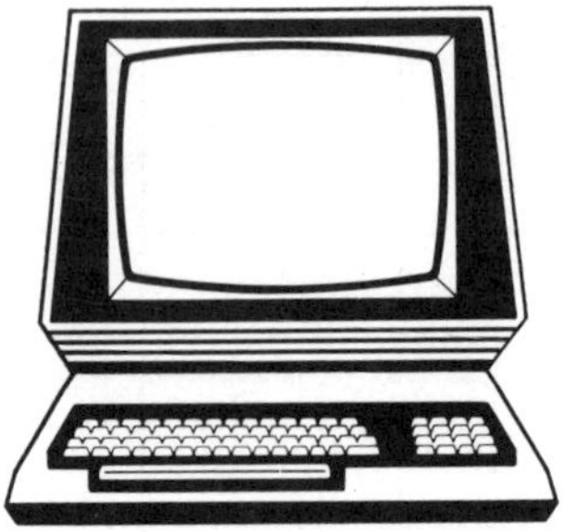

The Next Step

Now that you have completed the book and, hopefully, worked most of the projects, you have learned quite a bit about the use of ICs. You have used them in circuits and tested the circuits to gain an understanding of how the ICs perform and how they are used. The next step is to obtain some data books and application information either from the electronic stores or from the manufacturers of ICs. If you have studied this book and done the projects, you will be able to use the data and application books.

There are certain basic tools and equipment any serious electronic experimenter or hobbyist should have. These are required to construct the test electronic circuits. The following is a list of the basic tools required:

Small long nose pliers
Small dikes or side cutters
Small soldering iron
Small rosin core solder
Wire strippers
Small insulated wire, size 24 to 28
Test board, as described in Chapter 8
Multimeter
Power supply

These are the basic tools needed to get started. As circuits are built and projects completed, more tools will be added as you need them. For the soldering iron, choose a small 27 ½ watt iron with a small tip. This makes soldering to ICs and small components easier than with a large iron. Large irons have the tendency to burn components and insulation and to overheat the connections. When soldering, always keep a damp rag or sponge handy to wipe the soldering iron tip. After the iron has heated up, keep the tip covered with a layer of solder so that it is always shiny. This will make soldering considerably easier and is worth the extra effort.

Wire strippers are required to strip the insulation from the wires, and must be the proper size for the wire used. Using the wrong size strippers or using side cutters to strip the insulation usually ends up nicking the wire. This creates a weak spot in the wire and usually results in a broken wire. The multimeter measures voltage, current, and resistance. It does not have to be a digital multimeter, although digitals are nice. A good analog multimeter can be purchased for under twenty dollars. When assembling circuits, and when working with a circuit, always take time to make sure of every step. One misstep can result in component damage. Do not rush the assembly.

It must be expected that some circuits will not work as expected the first time. If the circuit does not work, remove the power and touch the components to see if they have overheated. Double check the circuit for wiring errors, shorts, or open connections. Most of the problems with electronic circuits are connection problems. Connection problems are broken wires, bad solder connections, loose connections, dirty contacts or connections, and the like. Patience is required to get some circuits working, but finding and correcting problems with circuits is a learning process.

The field of electronics is a fast growing and exciting field, with new ICs being introduced daily. This book just scratches the surface and is meant to serve as an introduction to ICs. From here, the opportunities are endless. In the past few years, the microprocessor and the single chip computer have become a reality. By taking it one step at a time, even the microcomputer can be mastered.

Index

OTHER POPULAR TAB BOOKS OF INTEREST